ENERGY Management in Illuminating Systems

ENERGY Management in Illuminating Systems

Kao Chen, P.E., FIEEE

CRC Press
Boca Raton London New York Washington, D.C.

Publisher: Robert Stern
Project Editor: Naomi Lynch
Marketing Managers: Barbara Glunn and Jane Stark
Cover design: Dawn Boyd

Library of Congress Cataloging-in-Publication Data

Chen, Kao, 1919–
Energy management in illuminating systems / Kao Chen.
p. cm.
Includes bibliographical references and index.
ISBN 0-8493-2628-1 (alk. paper)
1. Electric lighting—Energy consumption. 2. Electric power—Conservation. I.Title.
TK4169.C47 1999
621.32—dc21 99-20866
CIP

No claim to original U.S. Government works
International Standard Book Number 0-8493-2628-1
Library of Congress Card Number 99-20866
Printed in the United States of America 7 8 9 0
Printed on acid-free paper

Preface

The author is a power-major graduate. Following his graduation, he has had a career for more than 30 years at Westinghouse as a principal engineer, and then as a Fellow engineer in charge of power distribution and illuminating systems engineering and operations for 12 plants as well as several foreign operations. During the energy crunch years (mid 1970s to early 1980s) his important work was shifted to "relighting" operations. He developed a comprehensive relighting plan for the corporation and became a leader of the "Relighting Working Group." He emphasized that the primary purpose of relighting was to improve the workers' visual performance and thereby increase their productivity. However, the affiliated benefits of energy and cost savings will always be the secondary products of the program. Therefore, the program should begin with the determination of the most suitable lighting system for the workers' particular visual requirements. After the system is determined, the next step will be to select proper energy-efficient lamps, ballasts, and luminaires. Such an approach was proven to be much more technically viable than the simple replacement with greater efficiency units. In a period of 8 years, the team relighted over 100 plants with an annual energy savings of 90 million kWh and a cost savings of more than $3 million. For each relighting project the author followed up with field measurements and observations, and collected diagnostic data for the lamp and luminaire engineering colleagues. Together they made several significant improvements during the infancy period of energy-efficient lamps and luminaires.

With this accomplishment he was named the 1992 recipient of a coveted IEEE Richard Harold Kaufmann Award. During these years he was also a major contributor to the IEEE Standard Bronze Book, *Energy Management in Commercial and Industrial Facilities*. He presented many papers on relighting and related subjects at the IEEE-IAS annual conferences and received best committee and society paper awards nine times.

The author was elected an IEEE-Industry Applications Society's Distinguished Lecturer for 1996 and 1997. His lecture material was organized into three categories: (1) The Fundamentals of Energy Management; (2) Energy-Effective Illuminating Systems; and (3) The Impact of Energy-Efficient Equipment on System Power Quality. Subsequently these lectures were successfully delivered in five cities in China in 1996 and in four countries of Central America in 1997. The total combined audience for these two lecture tours easily exceeded 1000. They were practicing engineers and/or academicians representing the top experts in the field in each respective area. They showed great interest in these subjects, mainly because they have picked up some useful knowledge from the lectures which can be applied to their daily work. Some information in Chapter 9 is very new, such as the "K" factor for specifying transformers to handle nonlinear loads with heavy harmonics.

Therefore, the author feels that the material of his lectures is worthy of broader dissemination. What will be a better way of disseminating than publishing this book?

The book is organized along the above-mentioned categories of his lectures. The entire material is covered in ten chapters. Chapter 1, Introduction, brings out the main objectives of the book—to present the latest concept in energy management of illuminating systems. Chapter 2 presents the Fundamentals of Energy Management and emphasizes the triangular relationship of three ingredients—Organization, Economics, and Technology. It goes into great detail to discuss Time Value of Money and Economic Models for Evaluation. Chapter 3 extends applications of the energy management principles to illuminating systems and emphasizes an energy-effective system as the beginning of any successful energy management program. This leads to more detailed discussions in Chapters 4 to 6, which cover the new concepts in lighting design and proper choices of energy-efficient components—lamps, ballasts, and luminaires. Chapter 6 discusses the importance of how to incorporate daylight and apply properly selected controls to enhance energy savings. Chapter 7 covers many concrete examples of well-designed new installations as well as some ingenious retrofitting installations. It also highlights pitfalls of some retrofits and recommendations for their improvement. Chapter 8 covers methods of evaluating and achieving energy efficiency, and examples of life-cycle costing analysis. Chapter 9 discusses the impact of energy efficient equipment on system power quality, and goes into great detail on harmonics and introduces the "K" factor for specifying new transformers to handle nonlinear load with harmonics. Finally Chapter 10 summarizes the development of lighting energy standards and the effect of the recent EPAC (Energy Policy Act of 1992). The author wishes to point out that the essential difference of this book from many other books on similar subjects lies in the technical approach to a project, i.e., the emphasis on the sound engineering design and analysis of an illuminating system, whether new or retrofit.

It is hoped that practicing engineers and academicians all over the world will make good use of the information presented in the book. Successful implementation of energy management projects in the industrial and commercial facilities everywhere by these engineers and/or academicians will make the world a better and healthier environment in which to live in the next millennium.

My gratitude goes to Mr. Robert Stern, Ms. Naomi Lynch, Ms. Elizabeth Spangenberger, Ms. Felicia Shapiro, and Ms. Lyn Meany at CRC Press for their encouragement and support of this effort.

Kao Chen

Contents

chapter one

Introduction

1.1 The significance of energy management

Energy management deals with engineering, design, application, utilization, and, to some extent, the operation and maintenance of systems to promote the optimal use of electrical energy. "Optimal" in this case refers to the design or modification of a system to use minimum overall energy where the potential or real energy savings are justified on an economic or cost–benefit basis. Optimization also involves factors such as comfort, healthful working conditions, the practical aspects of productivity, the aesthetic acceptability of the space, and public relations.

Managing energy is more than the implementation of energy conservation opportunities. Since energy management is a continuous process, it is extremely important to monitor energy usage and to use the results to gauge future actions. In some energy applications, measurement or control, or both, at each point of application may be necessary to ensure efficient use of energy.

Accordingly, energy management involves the following professions and fields:

1. Engineering
2. Management
3. Economics
4. Financial analysis
5. System analysis
6. Public relations
7. Environmental engineering

Some of the essential tools for the program are

1. Meters and measurements
2. Demand and energy limits
3. Highly efficient energy devices
4. Control systems or building management systems

In engineering new facilities, potential energy conservation must be carefully carried out in initial design of various systems. For existing facilities energy conservation can also be achieved with properly engineered retrofits to the systems in service.

Once an energy-effective system has been established by either a new design or a successful retrofit, an energy management program can then be put into practice to (1) achieve optimal usage of energy, (2) deliver beneficial cost savings to the facility owners, and (3) indirectly reduce pollution and preserve the environment as its ultimate goal.

1.2 The role of energy-effective systems

Energy-effective systems are the most important ingredients of any successful energy management program. Component efficiency alone cannot create an energy-effective system, but technically sound system design and application of various efficient components can produce such a system. This is certainly true for the subject "Energy Management in Illuminating Systems." Ever since the energy crunch years lighting retrofits have been one of the most popular targets for energy conservation. However, based on numerous energy users' reports, too often the so-called successful projects are judged from their energy/cost savings or the payback years as the criteria; rarely a mention is made about improvements of visual environment as the result of retrofitting. This goal can usually be achieved along with energy/cost savings if the illumination design principles are properly observed and applied to a project in its initial stages.

An energy-effective illuminating system can be established by:

1. Recognizing the workers' visual task requirements and making appropriate design analysis of a proposed new or replacement system
2. Selecting best-suited energy-efficient light source and equipment for the proposed system
3. Optimizing the control techniques and integrating daylight into the design wherever feasible

In other words, the illuminating engineer must make sure that the system he/she designed will be energy-effective. Only the energy-effective systems can assure a successful energy management program in illuminating systems.

1.3 EPA's "Green Lights" and the 1992 National Energy Policy Act

The "Green Lights" (GL) program and the 1992 National Energy Policy Act are two recent important events which will affect illuminating systems design for years to come. All energy managers and/or illuminating design engineers should be familiar with the impacts or effects of these programs before they launch their energy-related illumination projects.

The GL program was introduced by the EPA (Environmental Protection Agency) in January 1991. Now more than 2000 corporate partners have joined in. The completely voluntary GL program encourages companies to use energy-efficient lighting products and techniques. These companies agree to survey all of their facilities and install new illuminating systems that maximize energy savings, are profitable, and do not compromise lighting quality in up to 90% of the space. They agree to complete the upgrades within five years, document their improvements, and also to use energy-efficient lighting products when designing new facilities. The EPA offers a training workshop which provides a lighting upgrade manual and a software program. The upgrade manual follows the same order used in a typical commercial building lighting retrofit: project planning, the building survey, financial options, lighting evaluation, project management, equipment disposal, lighting maintenance, and progress reporting. In this approach, it does not bypass expertise, i.e., the illuminating engineer or designer is still required to handle the project.

In October 1992, the President signed into law the most wide-sweeping, comprehensive energy legislation in recent history. The impact of this new energy policy will be felt by business and industry into the new century. Products affected by new efficiency standards include motors, furnaces, air conditioners, water heaters, showerheads, lighting fixtures, lamps, and transformers. The planned energy reduction will have immediate benefit of reducing environmental emissions of CO_2, SO_2, and a host of other chemicals. This will certainly reduce the "greenhouse effect," global warming, and acid rain.

In addition to the new lighting standard for new buildings (ASHRAE/IES 90.1), there will be a requirement for new efficiency standards to be met for the most popular types of lamps and lighting equipment. Detailed information on the effects of various types of fluorescent and incandescent lamps will be discussed in ensuing chapters. Effective October 1995, lamps, including 4-ft fluorescent, 2-ft U-shaped fluorescent, and incadescent reflector types, which do not meet energy standards are prohibited from further manufacturing. In short, the GL program and the EPACT tend to make the design of a new or retrofitting energy-effective illuminating system more complex and require an expert engineer to carry out the project, so that subsequent energy management programs can be properly implemented.

1.4 Revolution or evolution

The so-called energy-efficient lighting "revolution" sparkled by the National Energy Policy Act of 1992 and electric utilities' ongoing efforts to reduce power demand (so-called DSM) might better be called "Evolution." During the past year lighting manufacturers have brought to market both breakthrough technologies and new twists on older technologies to satisfy customers' particular and diverse needs. Spearheading this market are building owners and lighting specifiers who demand the most energy-efficient and effective lighting possible for each and every application.

A case in point is fiber-optic (FO) lighting technology which has long been used in Europe. FO lighting has now found a home in U.S. applications, such as light-sensitive museum exhibits. Using glass fiber eliminates ultraviolet and infra-red radiation from the spectrum without using heat filters. The induction lamp, another innovative product, is genuinely new technology introduced in 1994. Induction lamps are a great solution for applications that need long life, rugged construction, and energy efficiency. For London's Big Ben clock tower, the lamps offer brighter illumination and a projected life of 15 years. They are ideally suited. Another recent breakthrough is the "sulfur" lamp. It will move from the laboratory to the workplace with a solid-state microwave generator as its power supply. A central sulfur light source with FO distribution will soon be a reality. This new lighting technology outshines current systems in brightness and energy efficiency and offers a potentially limitless lifetime. It generates light by electromagnetically heating sulfur with microwave energy. The new lighting system consists of golf ball-sized bulbs connected to 10-ft-long, 10-in.-diameter light pipes. The system is ideal for illuminating large spaces, such as factories, warehouses, arenas, and shopping malls. Such a system is already being used at the Smithsonian's Air & Space Museum, where three 90-ft pipes have replaced 94 conventional lamps, boosted lighting level three times, and cut energy use by 25%. The lamp's life is limited only by the life of the microwave generator, which is approximately 10,000 to 15,000 h.

Energy savings is the driving force in the lighting industry. For new and retrofit lighting installations, the options include energy-efficient ballasts, high-efficiency lamps, and reflectors. In addition, the utilities offer a rebate to help reduce the initial cost and payback time for a relighting program.

1.5 Objectives of the book

Since the objectives of the book are far beyond achieving a successful energy management program in illuminating systems or an energy-effective illuminating system by virtue of proper design, it is intended to present the latest concept in energy management of illumination and how to implement the intended program to all engineers/managers who are responsible for carrying out their obligation to their facilities, industrial or commercial, for energy/cost savings and promoting and protecting the environment in which we live.

To summarize, an effective energy management program requires proper organization, conscientious controlling and monitoring, and a timely reporting system. However, to ensure the success of any energy management program, the illuminating system must be energy-effective to begin with.

This can be accomplished with sound engineering analysis and design principles carried out precisely during the early stage of the contemplated project. The ensuing chapters are arranged in the best sequence to help readers to achieve the above goal as envisioned by the author.

The ensuing chapters will first cover important terminologies in energy-efficient illuminating systems and energy management; energy-efficient lighting system components; new concepts and procedures in illumination design to achieve an energy-effective system which not only uses the latest energy-efficient components, but also integrates appropriate controls and daylighting techniques. Then there will be chapters on the retrofitting procedures and examples; evaluation of the system energy effectiveness and benefits. In addition, a whole chapter is devoted to the DSM, power quality, and harmonics which are generated from the application of energy-efficient lamps and electronic ballasts. The book ends with various updated lighting energy standards which are most relevant to today's illuminating systems design or retrofitting to realize energy optimization.

chapter two

Fundamentals of energy management

2.1 General concept

To recapitulate the discussions in the introduction, energy management, to be effective, must be based on three equally important elements, namely, technology, economics, and organization. The close relationship between each other or among the three may be more visually established as a triangle (Figure 2.1). In order to succeed in any energy management program, it is essential to firmly establish all requirements in these three branches.

2.2 Elements of energy management

2.2.1 Energy applications

To understand the energy consumption patterns in a facility, it is important to understand the applications of energy processes. Energy applications in a facility can be grouped into six major types:

1. Space conditioning — energy used directly for heating or cooling an area for comfort conditioning
2. Boiler fuel — this is subdivided into space conditioning and process energy
3. Direct process heat — energy used to heat the product being processed, e.g., for kilns, reheat furnaces, etc.
4. Feedstock — fuel used as an ingredient in the process, e.g., electroplating and sodium production
5. Lighting — new design of energy-effective lighting and/or retrofitting for energy-saving systems
6. Mechanical drive — motors used for ventilation system, pumps, crushers, grinders, production lines, etc.

Energy Management

Energy management is built on three ingredients:

Organization
Economics
Technology

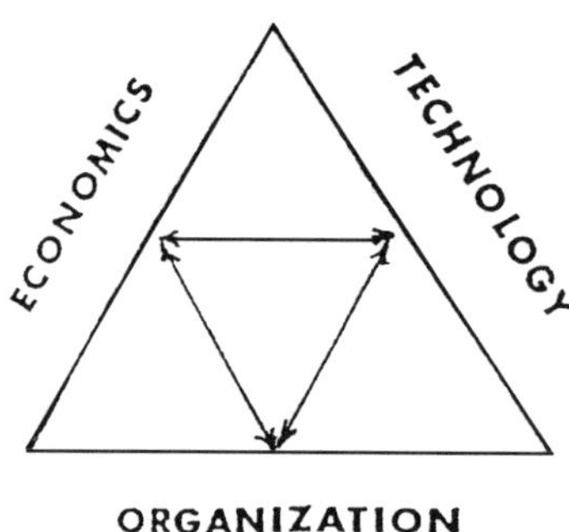

Figure 2.1 Three elements.

Energy savings can result from improvements in the efficiency of energy conversion processes. Any process requires a certain minimum consumption of energy. Energy additions beyond this minimum require an evaluation of the incremental cost of more efficient equipment or techniques vs. the resulting energy savings or costs. Some of the more intensive uses of industrial energy, including chemicals, paper, and petroleum refining, have long found it competitively advantageous to design for energy conservation. Virtually all new commercial/industrial facility designs consider energy savings in some form.

2.2.2 Energy-saving methods

In general, the following represent the most useful methods that can be applied:

1. Housekeeping measures — energy savings can result from better maintenance and operation. Such measures include shutting off unused equipment; improving electricity demand management; reducing winter temperature settings; turning off lights; and eliminating steam-compressed air and heat leaks. Proper cleaning and replacement of filters

in equipment, and periodic cleaning and lamp replacement in the lighting system will result in optimal energy use in existing facilities.
2. Equipment and process modifications — these can be either applied to existing equipment (retrofitting) or incorporated in the design of new equipment.
3. Better utilization of equipment — it can be achieved by carefully examining the production processes, schedules, and operating practices. Improvements in plant efficiency can be achieved through proper sequencing of process operations, rearranging schedules to utilize process equipment for continuous periods of operation to minimize losses associated with start-up; scheduling process operations during off-peak periods to level power demand; and conserving the use of energy during peak demand period. Commercial facilities will typically achieve energy savings by relamping, installing adjustable speed drives in ventilation systems, and considering solar effects.
4. Reduction of losses in the building shell — reduction in heat loss is achieved by adding insulation, closing doors, reducing exhaust, utilizing process heat, etc.

2.3 Organizing the program

2.3.1 Introduction

An energy management function gives the line managers the tools to get the job done. Line managers need to know their energy use and costs, the future energy supply availability and its cost, problems or opportunities of energy situations, and those alternative solutions worth pursuing.

At any level of the corporate structure, an individual should understand the incentives and motivations of top management. An alert engineer will become aware of the hidden lines of authority and the key persons who make decisions. It is the engineer's job to see that the proper facts are presented through the proper channels to convince top management that they should make the energy commitment.

The initial factors in organizing an effective energy program are

1. Obtain top management energy commitment — this is a formally communicated, financially supported dedication to reducing energy consumption while maintaining or improving the functioning of a facility.
2. Obtain people commitment — people at all levels of the organization should be involved in the program. Ideas should be encouraged with rewards for significant contributions to the energy management program.
3. Set up a communication channel — the purpose of this channel is to report to the organization the results of your efforts, to recognize high achievers, and to identify reward recipients.

4. Change or modify the organization to give authority and commensurate responsibility for the conservation effort and develop an energy management program.
5. Set up a means to monitor and control the program.

2.3.2 *Energy auditing*

The first step for the energy team is to use an energy audit to determine the amount of energy that enters and leaves a plant. This determination will probably be an approximation, but accuracy should improve with experience. The audit consists of a survey and appraisal for various site energy systems.

An audit will uncover such items as unnecessary operation of equipment, unnecessary high levels of lighting, no one assigned to turn off unnecessary lights and equipment, improper thermostat setting, unnecessary or excess ventilation, and other obvious waste. A comprehensive tour will also involve acquisition of nameplate data, motor and lighting loads, and other system details.

A comprehensive audit listing requires the development of a complete energy audit listing major energy-using equipment with nameplate data relating to energy, any efficiency tests, and estimates or measured hours of operation per month. This will include monthly utility data, amounts, and total costs. National Weather Service monthly degree days for heating and cooling must be included. Generally, a 1- or 2-year compilation of data is used. The crucial part is an intelligent appraisal of energy usage, what is done with energy as it flows through the processes and facility, and how this compares with accepted known standards. This study can lead to the development of a prioritized list of projects with a high rate of return.

2.3.3 *Energy monitoring and reporting*

Since energy management is a continuous process, it is extremely important to monitor energy usage and to use the results to gauge future actions. In some energy applications, such as in combustion processes, measurement or control (or both) at each point of application may be necessary to ensure efficient use of energy. Measuring methods that best describe the process are thus necessary to control, evaluate, and manage efforts to conserve energy. It is also important to budget the anticipated energy usage so that performance can be evaluated and plans can be made. A measure against budget also provides input to determine if additional action is required.

Several types of reports can be prepared to show energy usage over time. Summary reports for upper management typically note monthly or quarterly performance by facility or division. The type of report will vary with each specific facility. Some typical reports use the percent energy reduction method, the product energy rate method, and the activity method.

2.4 Technology

As discussed in Section 2.3.2, a walk-through audit consists of a careful observation of operation while walking through the facility with the express purpose of uncovering energy conservation opportunities (ECO). Six categories for energy use should be analyzed:

1. Lighting — the first item that gets attention is lighting because of its visibility. Before making changes, we must analyze and consider from conservation standpoint. The ensuing chapters of the book will concentrate on various energy aspects of an illuminating system, whether it be a new design or a retrofit.
2. HVAC (the heating, ventilation, and air conditioning) systems
3. Motors and drives
4. Process
5. Other electrical equipment (transformers, contactors, conductors, switchgears, ballasts, etc.)
6. Building shell (insulation and transmission)

Within each of these categories, the four basic energy-saving methods (refer to Section 2.2.2) should be considered. Since in all cases electric energy is required to perform work and make products, the engineer must be familiar with:

1. Utility rate structures
2. How to calculate the cost of electricity
3. How to evaluate loss — no load loss and load loss of various equipment
4. Demand control
5. DSM — demand-side management

2.5 Economics

The use of economic analysis is critical to the energy management program because monetary savings can significantly influence management decisions. The engineer should be able to translate a proposal into monetary value, i.e., expenditures vs. savings.

The engineer can use the economic basis to develop an energy program and to determine the best choice among alternatives. The determination should include understanding payback periods, the time value of money, and weighing other pertinent costs.

Economic models

There are two general means of evaluating energy options: simple break-even analysis and a more complex method called "life cycle costing" (LCC). When there is a large energy savings for a small investment, simple payback may be the best evaluating tool. Furthermore, housekeeping projects with minimal or no cost may not need evaluating. LCC is most likely needed when the project costs are large compared to the energy savings or when there are significant future costs.

1. Break-even analysis — simple payback analysis, break-even point analysis, and minimum payback analysis are essentially the same. The basic mathematical formula is

   Break-even point = net capital investment / savings per unit

   The break-even point has taken the name of payback from this minimum payback usage. Hence, the more common term is payback, and management is more likely to ask for the payback of a particular energy option. It is important to note that break-even analysis is not restricted to time as a base. Production and energy consumption are also good bases. For example, the break-even point can be calculated in terms of the amount of product manufactured such as dollar savings per pound of output.
2. Marginal cost analysis — marginal (or incremental) cost analysis is more a concept than an economic model. It is primarily used in calculating cogeneration sales to the utility and in the development of special rates or contracts. Marginal cost analysis is simply the determination and use of the next increment of the cost of money or cost of electric energy.
3. Life cycle costing — LCC is the evaluation of a proposal over a reasonable time period considering all pertinent costs and the time value of money. The evaluation can take the form of present value analysis or uniform annual cost analysis. The LCC method takes all costs and investments at their appropriate points in time and converts them to current costs. Inflation is assumed equal for all cost factors unless it is known to differ among cost items. Items for consideration include: (1) design cost, (2) initial investment, (3) overheads, (4) annual maintenance costs, (5) annual operating costs, (6) recurring costs, (7) energy costs, (8) salvage values, (9) economic life, (10) tax credits, (11) inflation, (12) cost of money.

2.6 Time value of money

The following are terms to be familiar with:

1. Annuity — a series of equal amounts evaluated at the end of equal time periods (usually 1 year) for a specific number of periods
2. Discount rate — the percentage rate used by a corporation that represents the time value of money for use in economic comparisons
3. Future worth (value) — the value of a sum of money at a future time

4. Present worth (value) — the value of an amount discounted to current dollars using the time value of money

In Table 2.1 — Time Value Factors — several factors are mathematically expressed for quick reference. Figure 2.2 is a time value chart to help to relate these factors.

Table 2.1 Time Value Factors

Symbol	Name	Description	Formula[a]
AP	Annuity payment	Equal amounts of money at the ends of a number of periods	—
FAF	Future worth annuity factor	Converts an annuity to an equivalent future amount	$\text{FAF} = \frac{(1+i)^n - 1}{i}$
FW	Future worth	The dollar amount (of an expense or investment) at a specific future time	—
FWF	Future worth factor	Converts a single present amount to an amount at a future point in time	$\text{FWF} = (1+i)^n$
PAF	Present worth annuity factor	Converts an annuity to a single present amount	$\text{PAF} = \frac{(1+i)^n - 1}{i(1+i)^n}$
PW	Present worth	The single value or worth today	—
PWF	Present worth factor	Converts a future amount to an amount today	$\text{PWF} = \frac{i}{(1+i)^n}$
SAF	Sinking fund annuity factor	Converts a future amount into an equivalent annuity	$\text{SAF} = \frac{i}{(1+i)^n - 1}$

a. The two variables have the following definitions: n = number of years in the evaluation period; i = interest rate or other cost of money factor used.

2.7 *Application of life-costing analysis*

Two examples of applications of LCC to practical engineering problems can support that the ultimate correct choice of options will be based on LCC, not the other economic models. The first example given here is for an engineer to make a most economical and correct choice of three options facing a failed motor. The second example deals with a choice of an economical and energy-efficient illuminating system which will be discussed in depth in Chapter 8.

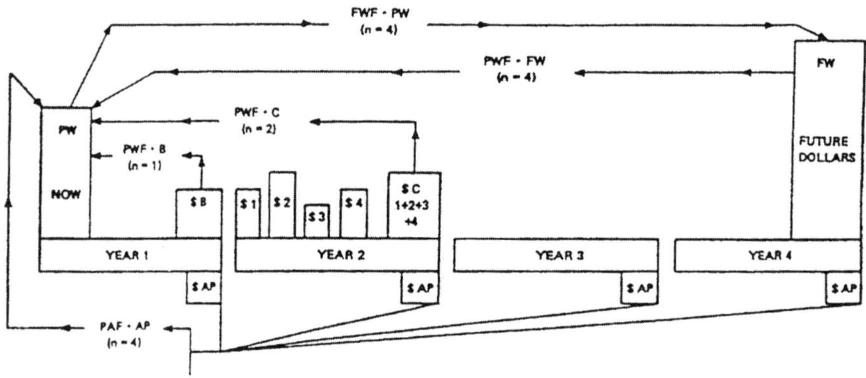

Figure 2.2 Time value chart.

Example 1

A 100-HP ODP motor which runs 5000 h/year at an operating load of 75% of design goes out of service. The engineer must decide:

1. Rewind the failed motor
2. Purchase a new motor of standard operating efficiency
3. Purchase a new motor of higher operating efficiency

- Option 1 — motor rewind, costs $1500 plus $200 for installation, results in a motor operating efficiency of 90.9% for the first year energy operating cost of $23,400, but will last for only 10 years
- Option 2 — new standard efficiency motor, costs $2100 plus $200 for installation, results in a motor operating efficiency of 91.9% for first year energy operating cost of $23,100, and will last for 15 years
- Option 3 — new high efficiency motor, costs $2500 plus $200 for installation, results in a motor operating efficiency of 94.8% for first year energy operating cost of $22,400, and will last for 15 years

In order to proceed with this problem, the following are several useful assumptions:

1. It is theoretically possible to rewind a motor with no loss in efficiency. However, for this problem we conservatively estimate that the effect of each of the three rewinds leads to a cumulative average efficiency decrease of 1%.
2. Assume that the cost of electricity is $0.06/kWh and $80/kW-year. The operating cost of Option 1 is derived as follows:

$$\text{hp} \times \text{loading} \times \text{kW/hp}/\text{ motor eff.} = \text{demand in kW}$$

3. We shall assume a 30-year investment horizon and that each option is exclusive of the other. The rate of inflation is assumed to be 4%/year and the corporate discount rate is set at 20%. The salvage value of each option at the end of 30 years is assumed to be equal for all three options, and so is not considered in this problem.
4. Finally, we do not consider the tax effects of the alternatives (such as depreciation).

Now all should be able to proceed with analysis and the LCC comparison for three options and come up with the correct answer.

Solutions using LCC analysis

Operating cost of Option 1:

$$(100 \text{ hp})\ (0.75 \text{ hp/hp}) \times 0.746\ (\text{kW/hp})\ /\ (0.909) = 61.55 \text{ kW}$$

$$(61.55 \text{ kW}) \times (80 \text{ kW/year}) = \$4{,}924$$

$$(61.55) \times (5000) \times (0.06) = \$18{,}465$$

$$\$4{,}924 + \$18{,}465 = \$23{,}389 \text{ (say, \$23,400)}$$

The LCC of Option 1 consists of three motor rewindings (in year 0, 10, and 20) and 30 years of operating costs.

The LCC of Option 2 consists of two standard efficiency motor purchases (in year 0 and 15) and 30 years of operating costs.

The LCC of Option 3 consists of two high efficiency motor purchases (in year 0 and 15) and 30 years of operating costs.

The present worth of the capital and installation costs of the future motor rewindings and future motor purchases is estimated using the future worth factors, FWF, to account for the effects of inflation; and present worth factors, PWF, to discount those future expenses into present-day amounts.

Present worth of capital and installation costs for **Option 1**:

The first rewind cost = \$1500 + \$200 = \$1700

FWF 4% inflation 10 years = $(1 + 0.04)^{10} = 1.480$

PWF 20% discount rate, 10 years = $1/(1 + 0.20)^{10} = 0.162$, so the second rewind cost = \$1700 × FWF × PWF = \$406

FWF 4% inflation 20 years = $(1 + 0.04)^{20} = 2.191$

PWF 20% discount rate, 20 years = $1/(1 + 0.20)^{20} = 0.026$, so the third rewind cost = \$1700 × FWF × PWF = \$97

Present worth of first, second, and third motor rewind = \$1700 + \$406 + \$97 = \$2203

Present worth of capital and installation costs for **Option 2**:
First standard efficiency motor cost = \$2100 + \$200 = \$2300

FWF 4% inflation 15 years = $(1 + 0.04)^{15} = 1.801$

PWF 20% discount rate, 15 years = $1/(1 + 0.20)^{15} = 0.065$, so the second standard efficiency motor cost = \$2300 × FWF × PWF = \$269

Present worth of first and second motor costs = \$2300 + \$269 = \$2569

Present worth of capital and installation costs of **Option 3**:
First high efficiency motor cost = \$2500 + \$200 = \$2700

FWF 4% inflation 15 years = $(1 + 0.04)^{15} = 1.801$

PWF 20% discount rate, 15 years = $1 / (1 + 0.20)^{15} = 0.065$, so the second high efficiency motor cost = \$2700 × FWF × PWF = \$316

Present worth of first and second motor costs = \$2700 + \$316 = \$3016

We define a **"real" discount rate** that is netted of inflation as follows:

$$\text{real discount rate} = \{(1 + i)/(1 + f)\} - 1$$

where i = cost of money used (the discount rate of 20%)
f = rate of inflation (in this case 4%)
real discount rate = $\{(1 + 0.20)/(1 + 0.04)\} - 1 = 0.154$

Present worth annuity factor for 30 years

$$= (1 + i^n) - 1/i(1 + i)^n$$

$$= \{(1 + 0.154)^{30} - 1\}/\{0.154(1 + 0.154)^{30}\}$$

$$= 6.411$$

Present worth of operating cost **Option 1**:

$$= \text{PAF} \times \text{annual operating costs}$$

$$= 6.411 \times 23{,}400 = \$150{,}017$$

Present worth of operating cost **Option 2**:

$$= 6.411 \times 23{,}100 = \$148{,}094$$

Present worth of operating cost **Option 3**:

$$= 6.411 \times 22{,}400 = \$143{,}606$$

Total LCC of each option is the sum of present worth of the capital and installation cost of each option and the present worth of operating cost of each option.

Total LCC for **Option 1** = $2203 + $150,017 = $152,220
Option 2 = $2569 + $148,094 = $150,663
Option 3 = $3016 + $143,606 = $146,622

Conclusion: Option 3 is the choice to proceed with.

chapter three

Energy management in illumination

3.1 Introduction

Electric lighting consumes a significant amount of energy. About 20 to 25% of all electricity used in buildings and about 5% of total energy consumption in the U.S. are used for illumination. Lighting also produces additional heat in buildings, which may be beneficial in cold climate, but generally this represents a significant load on air conditioning systems, the heat from lighting typically accounting for 15 to 20% of a building's cooling load.

Over the last decade, energy management has become a fundamental element of mainstream lighting design. There have been many factors contributing to this trend. These factors include utility programs for promoting demand-side management (DSM), the development and promulgation of energy codes and standards designed to minimize building energy waste, and ongoing technological development of lighting equipment to improve system efficacies and utilization efficiencies.

3.2 Illuminating systems design for energy efficiency

3.2.1 Illumination needs

The lighting needs of a space must be designed to optimize proper allocation and management of energy. Lighting needs may range from simple orientation to complex visual tasks. Important considerations for prolonged visual tasks associated with work environment include adequate illumination on the task surface, a proper balance of luminance between the task surface and surrounding horizontal and vertical surfaces, control of direct and reflected glare from all sources of light, and acceptable color rendering of task elements and surroundings. Other considerations may also apply, depending on details of the specific visual task performed.

When the nature and location of visual tasks can be identified, it is possible to reduce ambient illuminances and corresponding energy consumption by providing illumination more selectively. When specific tasks and their locations cannot be identified, a more uniform pattern of ambient illumination is generally provided, along with provisions for task lighting and local control. In applications where there are no prolonged visual tasks, lighting for emphasis, aesthetics, and safety is of prime consideration.

3.2.2 *Space design and utilization*

Space design and utilization characteristics are often determined before the illuminating systems are considered. However, it will be necessary to coordinate design of illuminating and control systems with these characteristics in order to maximize the energy efficiency potential. For large interior space, it should be 0.70 or higher. This will increase the brightness within the space and will reduce the amount of electric lighting needed to produce a given illuminance.

The design and utilization of a space may also facilitate effective use of daylight. High reflectance interior finishes can enhance the effectiveness of daylighting for offsetting electric lighting.

Applications that involve similar visual tasks should be grouped together to optimize the energy usage for lighting. Using a base level of ambient illumination with different tasks, specific luminaires provide an additional degree of flexibility and economy by allowing luminaires to be easily relocated when the function of a space changes.

Occupancy schedules of a space should be planned ahead to optimize the effectiveness of lighting controls. Traffic patterns should also be determined so that local applications for occupancy sensors can be identified. This usually requires coordination among members of the project team.

3.2.3 *Lighting controls and daylighting*

Lighting controls — The most efficient luminaires and light sources can be utilized more effectively by integrating them with lighting controls that make the illumination more responsive to changing requirements. Lighting control strategies should be implemented either centrally over an entire building or locally in individual areas. Various combinations of these two levels of control are also commonly applied.

Central lighting control systems allow for lighting load scheduling to reduce peak demand as well as provide the ability to monitor and control total energy use. Central lighting control systems may be interfaced with the overall building management system (BMS). Most energy codes require that discrete spaces provide some means for controlling lighting within that space. A building may be divided into areas that require different lighting needs and different control strategies. Private zones consisting mainly of private offices should be differentiated from core zones utilizing an open-plan landscape. Private offices are good candidates for automated lighting control because they have predictable occupancy schedules. They often utilize a significant daylight contribution.

Lighting control strategies are closely related to space utilization. It is important to understand and coordinate space utilization in order to optimize the effectiveness of lighting controls for energy management. The system should be flexible so that a minimum number of luminaires are used at night or on weekends. Care should be taken that off-hour operation does not defeat automatic shut-off controls, allowing lighting to operate all night or all weekend. Detailed discussions of lighting controls will be given in Chapter 5.

Daylighting — Daylight can be an excellent source of ambient illumination. The potential for daylight utilization should be evaluated early in the design development of a space. Architectural features such as overhangs, light shelves, and window treatments may be incorporated into the design to enhance daylight utilization and control. For effective use of daylight in energy management, the levels and hours of daylight availability must be determined. The manner in which daylight is distributed in the space is important. Glare from fenestration should be controlled to the same degree as glare from luminaires. The best daylighting designs maximize daylight penetration into the space while minimizing the negative effects of direct sun.

The heat gain or loss through fenestration should be coordinated with the building envelope and HVAC systems. The use of exterior sun control devices and "high-performance" heat reflecting and insulating glass in windows should be considered to minimize solar heat gain in the summer and heat loss in the winter without obstructing views of the exterior. Daylighting reduces energy consumption if electric lighting can be reduced. Therefore, the electric lighting design should be coordinated with available daylight so that illuminance and distribution are integrated. Automatic dimming controls should be considered for continuous adjustment of electric lighting levels to achieve maximum energy savings and occupant acceptance.

More detailed discussions on daylighting will also be given in Chapter 5.

3.2.4 Light sources and luminaires

Electric light sources should be selected to maintain the highest efficacy, while providing proper color qualities, physical and optical size, and long-life operating characteristics (warm-up time, restrike, dimming, etc.). These attributes are related to decisions on luminaire types, lighting controls, and the general operation and maintenance after installation.

Table 3.1 shows the relative efficacy ranges of commonly used light sources, including ballast losses. Within a range, the higher-wattage sources are generally more efficient than the lower-wattage sources. High-pressure sodium, metal halide, and fluorescent are the most efficient white light sources; mercury vapor and standard incandescent are the least efficient.

Table 3.1 Efficacies of Various Light Sources

	Approximate lumens per watt
Candle (luminous efficacy equivalent)	0.1
Oil lamp (luminous efficacy equivalent)	0.3
Orignial incandescent lamp (1879)	1.4
60-W Carbon filament lamp (1905)	4.0
60-W Coiled coil tungsten filament lamp (1968)	14.7
1000-W General service lamp (1961)	23.0
No. 1 photoflood lamp (1961)	34.6
400-W Mercury-fluorescent lamp, deluxe white (1968)	57.5[a]
40-W Fluorescent lamp, cool white (1968)	80.0[a]
96-in. High output fluorescent lamp, white (1968)	82.0[a]
400-W Metal halide lamp	85.0[a]
1000-W Metal halide lamp	100.0[a]
400-W High-pressure sodium lamp	125.0[a]
1000-W High-pressure sodium lamp	130.0[a]
180-W Low-pressure sodium lamp	183.0[a]

a. Lamp only.

Except for incandescent- and halogen-type lamps, each light source requires a specific ballast. Ballasts efficiencies also vary and can have an effect on the total illuminating system efficiency.

Fluorescent electronic ballasts are significantly more efficient than their conventional magnetic counterparts and generally improve lighting quality by reducing lamp flicker and ballast hum.

There are other issues that must be considered when selecting a ballast, One of these is "ballast factor", which represents the percentage of rated lamp lumens that will be produced by a specific lamp/ballast combination. Some ballast types are available as "low light output," with a low ballast factor to provide the option of lowering light output and proportionally reducing energy consumption. A method for comparing the efficiency of lamp/ballast systems on a normalized basis is the calculation of ballast efficiency factor (BEF). BEF is a ratio of ballast factor (BF) divided by ballast input power (BIP).

$$BEF = BF/BIP$$

A higher BEF value indicates a more efficient lamp/ballast system. Chapter 6 provides detailed information pertaining to the operation and performance of various lamp types. Lamp manufacturers' published data provide specific information relative to lumen output, efficacy, life expectancy, lumen maintenance, and costs. Chapter 6 also covers information pertaining to various types of ballasts, including electronic types.

Luminaires — A luminaire is an assembly of individual components including lamps, ballasts, sockets, and wiring, and optical media such as reflectors, louvers, and lenses. The efficiency of the luminaire is affected by the performance of these individual components. Typically, luminaire configurations are constructed to perform specific functions for specific applications. Application constraints such as ambient temperature, color requirements, accessibility, and glare control needs may require the use of certain components or preclude the use of others. The efficiency of a luminaire must be evaluated relative to specific application criteria.

Luminaires that can be cleaned easily and those with low dirt accumulation will remain greater efficiency and reduce maintenance costs over the system life. When possible, luminaires with heat transfer capabilities should be considered for interior applications, so that heat generated by the lighting system can be effectively utilized or removed from a space coordinated with the overall building HVAC design.

Data from luminaire manufacturers are useful for determining how efficiently luminaires meet lighting needs. The efficiency of a luminaire is the percentage of lamp lumens produced that actually exits from the luminaire. Efficiency is simply a quantitative metric of photometric performance and does not provide any indication of lighting quality. More detailed information on various types of luminaires will be covered in Chapter 6.

3.3 *Energy efficiency for existing buildings*

In existing buildings, it is desirable that the lighting comply with the same energy standards as in new buildings. Improved energy utilization options include modifying or replacing the lighting system with a more efficient one, using reduced-wattage sources in spaces which are overlit, and modifying the operating characteristics of the building to reduce hours of use. Energy management strategies for new buildings are covered in this book elsewhere, as well as techniques and tools for surveying and evaluating existing lighting systems.

An existing building is evaluated from the standpoint of power and energy by the following steps:

1. Building survey and assessment
2. Power budget and limit determination
3. Energy limit determination
4. Energy limit analysis

Building Survey

The lighting systems of a building are assessed to determine the existing conditions of lighting, current lighting needs, the connected electrical load, hours of operation, and the connected control system. The electrical load of all connected luminaires, including lamp watts, ballast watts, and losses introduced by dimming devices, should be documented. The power for portable and supplementary lighting devices should be included in the total. Factors related to occupancy, daylighting, and controls should be included in the energy limit analysis. Specific information on surface finishes and reflectances should also be recorded. Sampling procedures have been used to avoid measuring every feature of the space.

Lighting power budget and energy limit determination

By using the lighting power density (unit power density procedure expressed in watts/square foot) in local energy codes or ASHRAE/IES 90.1-1989, the budgets for individual spaces and the limit for the building are calculated.

Beginning with the power budgets for each space, determined by the UPD procedure, energy budgets are determined by adjusting for occupancy schedules and the use of daylight and lighting controls.

The actual current use for the facility is determined and compared with the energy limit. The difference is the energy savings from the retrofits. The resulting revised energy use estimate can be compared with the recommended energy limit. If it exceeds the limit, the lighting systems should be reevaluated. Spaces or strategies which are inefficient should be identified and modified to reduce the estimated energy below the calculated limit.

Reference LEM-3-87 contains the full procedure and energy management forms to estimate and evaluate lighting energy use for buildings.

3.4 Lighting efficiency standards

Lighting energy efficiency standards are generally designed to establish minimum levels of energy management. This may be accomplished by setting minimum efficiency requirements for specific components, or by limiting the amount of power available for lighting, while still enabling the designer to provide quality lighting for task and spatial requirements of occupants. A saving in energy (kWh) can be achieved by reducing either the amount of connected load for the lighting system, or the time the lighting system is operating. The overall goal is to minimize energy consumed by a lighting system without compromising the quality of the lighting.

Two approaches are in common use for regulating lighting energy use:

1. The use of application standards which limit the total power allowed for the lighting system
2. Equipment regulations which mandate minimum efficiencies for system components

Over the last decade, both application standards and equipment regulations have been enacted into building energy codes by federal, state, and provincial governments in North America.

Chapter 10 will cover more detailed information on these subjects.

References

ASHRAE/IESNA 90.1-89 Energy Efficient Design of New Buildings, IESNA, New York, NY 10005-4001.

LEM-3-87 Design Considerations for Effective Building Lighting Energy Utilization, IESNA, New York, NY 10005-4001.

chapter four

New concepts and procedures in illumination design

4.1 *Determination of illumination levels*

Among the many new concepts for illumination design, the first to be discussed is the new method of determining illuminance levels. In the past when illuminating engineers wanted to find the recommended illuminance level for a given task, they would look in the lighting handbook to find a recommended level and then design an illuminating system for the task using the value as a minimum. This procedure provides very little latitude for fine-tuning an illumination design. In the new methods, a more comprehensive investigation of required illuminance is performed according to the following steps:

1. Instead of a single recommended illuminance value, a category letter is assigned. Table 4.1 shows different category letters for a selected group of industries.
2. The category letters are used to define a range of illuminance. Table 4.2 details illuminance categories and illuminance values for generic types of activities in interiors.
3. From within the recommended range of illuminance, a specific value of illuminance is selected after consideration is given to the average age of workers, the importance of speed and accuracy, and the reflectance of task background.

The importance of acknowledging the speed and accuracy with which a task must be performed is readily recognized. Less obvious is the need to consider the age of workers and the reflectance of task background. To compensate for reduced visual acuity, more illuminance is needed. Using the average age of workers as the age criterion is a compromise between the need of the young and older workers and, therefore, a valid criterion.

Table 4.1 Illuminance Categories for Selected Group of Industries

Area/Activity	Illuminance Category
Aircraft maintenance	a
Aircraft manufacturing	a
Assembly	
Simple	D
Moderately difficult	E
Difficult	F
Very difficult	G
Exacting	H
Automobile manufacturing	a
Bakeries	
Mixing room	D
Face of shelves	D
Inside of mixing bowl	D
Fermentation room	D
Make-up room	
Bread	D
Sweet yeast-raised products	D
Proofing room	D
Oven room	D
Fillings and other ingredients	D
Decorating and icing	
Mechanical	D
Hand	E
Scales and thermometers	D
Wrapping	D
Book binding	
Folding, assembling, pasting	D
Cutting, punching, stitching	E
Embossing and inspection	F
Breweries	
Brew house	D
Boiling and keg washing	D
Filling (bottles, cans, kegs)	D
Candy making	
Box department	D
Chocolate department	
Husking, winnowing, fat extraction, crushing and refining, feeding	D
Bean cleaning, sorting, dipping, packing, wrapping	D
Milling	E
Cream making	
Mixing, cooking, molding	E
Gum drops and jellied forms	D
Hand decorating	D
Hard candy	
Mixing, cooking, molding	D
Die cutting and sorting	E
Kiss making and wrapping	E
Canning and preserving	
Initial grading raw material samples	D
Tomatoes	E
Color grading and cutting rooms	F
Preparation	
Preliminary sorting	
Apricots and peaches	D
Tomatoes	E
Olives	F
Cutting and pitting	E
Final sorting	E
Canning	
Continuous-belt canning	E
Sink canning	E
Hand packing	D
Olives	E
Examination of canned samples	F
Container handling	
Inspection	F
Can unscramblers	E
Labeling and cartoning	D
Casting (see Foundries)	
Central stations (see Electric generating stations)	
Chemical plants (see Petroleum and chemical plants)	
Clay and concrete products	
Grinding, filter presses, kiln rooms	C
Molding, pressing, cleaning, trimming	D
Enameling	E
Color and glazing – rough work	E
Color and glazing – fine work	F
Cleaning and pressing industry	
Checking and sorting	E
Dry and wet cleaning and steaming	E
Inspection and spotting	G
Pressing	F
Repair and alteration	F
Cloth products	
Cloth inspection	I
Cutting	G
Sewing	G
Pressing	F
Clothing manufacture (see Sewn Products)	
Receiving, opening, storing, shipping	D
Examining (perching)	I
Sponging, decating, winding, measuring	D
Piling up and marking	E
Cutting	G
Pattern making, preparation of trimming, piping, canvas and shoulder pads	E
Fitting, bundling, shading, stitching	D
Shops	F
Inspection	G
Pressing	F
Sewing	G
Control rooms (see Electric generating stations – interior)	
Corridors (see Service spaces)	
Cotton gin industry	
Overhead equipment – separators, driers, grid cleaners, stick machines, conveyers, feeders and catwalks	D
Gin stand	D
Control console	D
Lint cleaner	D
Bale press	D
Dairy farms (see Farms)	
Dairy products	
Fluid milk industry	
Boiler room	D
Bottle storage	D
Bottle sorting	E
Bottle washers	b
Can washers	D
Cooling equipment	D
Filing: inspection	E
Gauges (on face)	E
Laboratories	E
Meter panels (on face)	E
Pasteurizers	D
Separators	D
Storage refrigerator	D
Tanks, vats	
Light interiors	C
Dark interiors	E
Thermometer (on face)	E
Weighing room	D
Scales	E
Dispatch boards (see Electric generating stations – interior)	
Electrical equipment manufacturing	
Impregnating	D

Table 4.1 (continued) Illuminance Categories for Selected Group of Industries

Area/Activity	Illuminance Category
Insulating: coil winding	E
Electric generating stations – interior (see also Nuclear power plants	
Air-conditioning equipment, air preheater and fan floor, ash sluicing	B
Auxiliaries, pumps, tanks, compressors, gauge area	C
Battery rooms	D
Boiler platforms	C
Cable room	B
Coal handling systems	B
Coal pulverizer	C
Condensers, deaerator floor, evaporator floor, heater floors	B
Control rooms	
Main control boards	D[c]
Auxiliary control panels	D[c]
Operator's station	E[c]
Maintenance and wiring areas	D
Emergency operating lighting	C
Gauge reading	D
Hydrocarbon and carbon dioxide manifold area	C
Laboratory	E
Precipitators	B
Screen house	C
Soot or slag blower platform	C
Steam headers and throttles	B
Switchgear and motor control centers	D
Telephone and communication equipment rooms	D
Tunnels or galleries, piping and electrical	B
Turbine building	
Operating floor	D
Below operating floor	C
Visitor's gallery	C
Water treating area	D
Elevators (see Service spaces)	
Explosives manufacturing	
Hand furnaces, boiling tanks, stationary driers, stationary and gravity crystalizers	D
Mechanical furnace, generators and stills, mechanical driers, evaporators, filtration, mechanical crystallizers	
Tanks for cooking, extractors, percolators, nitrators	D
Farms – dairy	
Milking operation area (milking parlor and stall barn)	
General	C
Cow's udder	D
Milk handling equipment and storage area (milk house or milk room)	
General	C
Washing area	E
Bulk tank interior	E
Loading platform	C
Feeding area (stall barn feed alley, pens, loose housing feed area)	C
Feed storage area – forage	
Haymow	A
Hay inspection area	C
Ladders and stairs	C
Silo	A
Silo room	C
Feed storage area – grain and concentrate	
Grain bin	A
Concentrate storage area	B
Feed processing area	B
Livestock housing area (community, maternity, individual calf pens, loose housing holding and resting areas)	B
Machine storage area (garage and machine shed)	B
Farm shop area	
Active storage area	B
General shop area (machinery repair, rough sawing)	D
Rough bench and machine work (painting, fine storage, ordinary sheet metal work, welding, medium benchwork)	D
Medium bench and machine work (fine woodworking, drill press, metal lathe, grinder)	E
Miscellaneous areas	
Farm office	
Restrooms (see Service spaces)	
Pumphouse	C
Farms – poultry (see Poultry Industry)	
Flour mills	
Rolling, sifting, purifying	E
Packing	D
Product control	F
Cleaning, screens, man lifts, aisleways and walkways, bin checking	D
Forge shops	E
Foundries	
Annealing (furnaces)	D
Cleaning	D
Core making	
Fine	F
Medium	E
Grinding and chipping	F
Inspection	
Fine	G
Medium	F
Molding	
Medium	F
Large	E
Pouring	E
Sorting	E
Cupola	C
Shakeout	D
Garages – parking	
Garages – service	
Repairs	E
Active traffic areas	C
Write-up	D
Glass works	
Mix and furnace rooms, pressing and lehr, glassblowing machines	C
Grinding, cutting, silvering	D
Fine grinding, beveling, polishing	E
Inspection, etching and decorating	F
Glove manufacturing (see Sewn Products)	
Hangars (see Aircraft manufacturing)	
Hat manufacturing	
Dyeing, stiffening, braiding, cleaning, refining	E
Forming, sizing, pouncing, flanging, finishing, ironing	F
Sewing	G
Inspection	
Simple	D
Moderately difficult	E
Difficult	F
Very difficult	G
Exacting	H
Iron and steel manufacturing	G
Laundries	
Washing	D
Flat work ironing, weighing, listing, marking	D
Machine and press finishing, sorting	E
Fine hand ironing	E
Leather manufacturing	
Cleaning, tanning and stretching, vats	D
Cutting, fleshing and stuffing	D
Finishing and scarfing	E

Table 4.1 (continued) Illuminance Categories for Selected Group of Industries

Area/Activity	Illuminance Category
Leather working	
Pressing, winding, glazing	F
Grading, matching, cutting, scarfing, sewing	G
Locker rooms	C
Machine shops	
Rough bench or machine work	D
Medium bench or machine work, ordinary automatic machines, rough grinding, medium buffing and polishing	E
Fine bench or machine work, fine automatic machines, medium grinding, fine buffing and polishing	G
Extra-fine bench or machine work, grinding, fine work	H
Materials handling	
Wrapping, packing, labeling	D
Picking stock, classifying	D
Loading, inside truck bodies and freight cars	C
Meat packing	
Slaughtering	D
Cleaning, cutting, cooking, grinding, canning, packing	D
Nuclear power plants (see also Electric generating stations)	
Auxiliary building, uncontrolled access areas	C
Controlled access areas	
Count room	E[c]
Laboratory	E
Health physics office	F
Medical aid room	F
Hot laundry	D
Storage room	C
Engineered safety features equipment	D
Diesel generator building	D
Fuel handling building	
Operating floor	D
Below operating floor	C
Off gas building	C
Radwaste building	D
Reactor building	
Operating floor	D
Below operating floor	C
Offices	
Packing and boxing (see Materials handling)	
Paint manufacturing	
Processing	D
Mix comparison	F
Paint shops	
Dipping, simple spraying, firing	D
Rubbing, ordinary hand painting and finishing art, stencil and special spraying	D
Fine hand painting and finishing	E
Extra-fine hand painting and finishing	G
Paper-box manufacturing	E
Paper manufacturing	
Beaters, grinding, calendering	D
Finishing, cutting, trimming, papermaking machines	E
Hand counting, wet end of paper machine	E
Paper machine reel, paper inspection, and laboratories	F
Rewinder	F
Petroleum and chemical plants	a
Plating	D
Polishing and burnishing (see Machine shops)	
Power plants (see Electric generating stations)	
Poultry industry (see also Farm – dairy)	
Brooding, production, and laying houses	
Feeding, inspection, cleaning	C
Charts and records	D
Thermometers, thermostats, time clocks	D
Hatcheries	
General area and loading platform	C
Inside incubators	D
Dubbing station	F
Sexing	H
Egg handling, packing, and shipping	
General cleanliness	E
Egg quality inspection	E
Loading platform, egg storage area, etc.	C
Egg processing	
General lighting	E
Fowl processing plant	
General (excluding killing and unloading area)	E
Government inspection station and grading stations	E
Unloading and killing area	C
Feed storage	
Grain, feed rations	C
Processing	C
Charts and records	D
Machine storage area (garage and machine shed)	B
Printing industries	
Type foundries	
Matrix making, dressing type	E
Font assembly – sorting	D
Casting	E
Printing plants	
Color inspection and appraisal	F
Machine composition	E
Composing room	E
Presses	E
Imposing stones	F
Proofreading	F
Electrotyping	
Molding, routing, finishing, leveling molds, trimming	E
Blocking, tinning	D
Electroplating, washing, backing	D
Photoengraving	
Etching, staging, blocking	D
Routing, finishing, proofing	E
Tint laying, masking	E
Quality Control (see Inspection)	
Receiving and shipping (see Materials handling)	
Rubber goods – mechanical	a
Rubber tire manufacturing	a
Safety	
Sawmills	
Secondary log deck	B
Head saw (cutting area viewed by sawyer)	E
Head saw outfeed	B
Machine in-feeds (bull edger, resaws, edgers, trim, hula saws, planers)	B
Main mill floor (base lighting)	A
Sorting tables	D
Rough lumber grading	D
Finished lumber grading	F
Dry lumber warehouse (planer)	C
Dry kiln coiling shed	B
Chipper infeed	B
Basement areas	
Active	A
Inactive	A
Filing room (work areas)	E
Service spaces (see also Storage rooms)	
Stairways, corridors	B
Elevators, freight and passenger	B
Toilets and wash rooms	C

Table 4.1 (continued) Illuminance Categories for Selected Group of Industries

Area/Activity	Illuminance Category
Sewn products	
Receiving, packing, shipping	E
Opening, raw goods storage	E
Designing, pattern-drafting, pattern grading and markermaking	F
Computerized designing, pattern-making and grading, digitizing, marker-making, and plotting	B
Cloth inspection and perching	I
Spreading and cutting (includes computerized cutting)	F[g]
Fitting, sorting and blunding, shading, stitch marking	G
Sewing	G
Pressing	F
In-process and final inspection	G
Finished goods storage and picking orders	F[h]
Trim preparation, piping, canvas and shoulder pads	F
Machine repair shops	G
Knitting	F
Sponging, decating, rewinding, measuring	E
Hat manufacture (see Hat Manufacture)	
Leather working (see Leather Working)	
Shoe manufacturing (see Shoe Manufacturing	
Sheet metal works	
Miscellaneous machines, ordinary bench work	E
Presses, shears, stamps, spinning, medium bench work	E
Punches	E
Tin plate inspection, galvanized	F
Scribing	F
Shoe manufacturing – leather	
Cutting and stitching	
Cutting tables	G
Marking, buttonholing, skiving, sorting, vamping, counting	G
Stitching, dark materials	G
Making and finishing, nailers, sole layers, welt beaters and scarfers, trimmers, welters, lasters, edge setters, sluggers, randers, wheelers, treers, cleaning, spraying, buffing, polishing, embossing	F
Shoe manufacturing – rubber	
Washing, coating, mill run compounding	D
Varnishing, vulcanizing, calendering, upper and sole cutting	D
Sole rolling, lining, making and finishing processes	E
Soap manufacturing	
Kettle houses, cutting, soap chip and powder	D
Stamping, wrapping and packing, filing and packing soap powder	D
Stairways (see Service spaces)	
Steel (see Iron and Steel)	
Storage battery manufacturing	D
Storage rooms or warehouses	
Inactive	B
Active	
Rough, bulky items	C
Small items	D
Structural steel fabrication	E
Sugar refining	
Grading	E
Color inspection	F
Testing	
General	D
Exacting tests, extra-fine instruments, scales, etc.	F
Textile mills	
Staple fiber preparation	
Stock dyeing, tinting	D
Sorting and grading (wood and cotton)	E[d]
Yarn manufacturing	
Opening and picking (chute feed)	D
Carding (nonwoven web formation)	D[e]
Drawing (gilling, pin drafting)	D
Combing	D[e]
Roving (slubbing, fly frame)	E
Spinning (cab spinning, twisting, texturing)	E
Yarn preparation	
Winding, quilling, twisting	E
Warping (beaming, sizing)	F[d]
Warp tie-in or drawing-in automatic)	E
Fabric production	
Weaving, knitting, tufting	F
Inspection	G[d]
Finishing	
Fabric preparation (desizing, sourcing, bleaching, singeing, and mercerization)	D
Fabric dyeing (printing)	D
Fabric finishing (calendaring, sanforizing, sueding, chemical treatment)	E[d]
Inspection	G[d,f]
Tobacco products	
Drying, stripping	D
Grading and sorting	F
Toilets and wash rooms (see Service spaces)	
Upholstering	F
Warehouse (see Storage rooms)	
Welding	
Orientation	D
Precision manual arc-welding	H
Woodworking	
Rough sawing and bench work	D
Sizing, planing, rough sanding, medium quality machine and bench work, gluing, veneering, cooperage	D
Fine bench and machine work, fine sanding and finishing	E

a Industry representatives have established a table of single illuminance values which, in their opinion, can be used. Illuminance values for specific operations can also be determined using illuminance categories of similar tasks and activities found in this table and the application of the appropriate weighting factors in Table 4.3 and 4.4.

b Special lighting such that (1) the luminous area is large enough to cover the surface which is being inspected and (2) the luminance is within the limits necessary to obtain comfortable contrast conditions. This involves the use of sources of large area and relatively low luminance in which the source luminance is the principal factor rather than the illuminance produced at a given point.

c Maximum levels – controlled system.

d Supplementary lighting should be provided in this space to produce the higher levels required for specific seeing tasks involved.

e Additional lighting needs to be provided for maintenance only.

f Color temperature of the light source is important for color matching.

g Higher levels from local lighting may be required for manually operated cutting machines.

h If color matching is critical, use illuminance category G.

Table 4.2 Illuminance Categories and Illuminance Values for Generic Types of Activities in Interiors

Type of activity	Illuminance category	Ranges of illuminances: Lux	Ranges of illuminances: Footcandle	Reference work-plane
Public spaces with dark surroundings	A	20–30–-50	2–3–5	
Simple orientation for short temporary visits	B	50–75–100	5–7.5–10	General lighting throughout spaces
Working spaces where visual tasks are only occasionally performed	C	100–150–200	10–15–20	
Performance of visual tasks of high contrast or large size	D	200–300–500	20–30–50	
Performance of visual tasks of medium contrast or small size	E	500–750–1000	50–75–100	Illuminance on task
Performance of visual tasks of low contrast or very small size	F	1000–1500–2000	100–150–200	
Performance of visual tasks of low contrast and very small size over a prolonged period	G	2000–3000–5000	200–300–500	Illuminance on task obtained by a combination of general and local (supplementary lighting)
Performance of very prolonged and exacting visual tasks	H	5000–7500–10000	500–750–1000	
Performance of very special visual tasks of extremely low contrast and small size	I	10000–15000–20000	1000–1500–2000	

Task background affects the ability to see because it affects contrast, an important aspect of visibility. More illuminance is required to enhance the visibility of tasks with poor contrast. Reflectance is calculated by dividing the reflected value by the incident value. The data given in Tables 4.3 and 4.4 are taken from the *IES Lighting Handbook* and are applied to provide a single value of illuminance from within the range recommended.

Table 4.3 Weighting Factors for Selecting Specific Illuminance Within Ranges A, B, and C

Occupant and room characteristics	Weighting factor		
	–1	0	+1
Worker's age (average)	Under 40	40–55	Over 55
Average room reflectance [a]	> 70%	30–70%	< 30%

Source: *IES Lighting Handbook*, 6th ed.

Note: This table is used for assessing weighting factors in rooms where a task is not involved.

1. Assign the appropriate weighting factor for each characteristic.
2. Add the two weights; refer to Table 4.2, Categories A through C:
   a. If the algebraic sum is –1 or –2, use the lowest range value.
   b. If the algebraic sum is 0, use the middle range value.
   c. If the algebraic sum is +1 or +2, use the hightest range value.

a. To obtain average room reflectance: determine the areas of ceiling, walls, and floor; add the three to establish room surface area; deterine the proportion of each surface area to the total; multiply each proportion by the pertinent surface reflectance; and add the three numbers obtained.

4.2 Illumination computational methods

4.2.1 Zonal cavity method

Introduced in 1964, the zonal cavity method of performing lighting computations has gained rapid acceptance as the preferred way to calculate number and placement of luminaires required to satisfy a specified illuminance level requirement. Zonal cavity provides a higher degree of accuracy than does the old lumen method, because it gives individual consideration to factors that are glossed over empirically in the lumen method.

Definition of cavities — With the zonal cavity method, the room is considered to contain three vertical zones of cavities. Figure 4.1 defines the various cavities used in this method of computation. Height for luminaire to ceiling is designated as the ceiling cavity (h_{cc}). Distance from luminaire to the work plane is the room cavity (h_{rc}), and the floor cavity (h_{fc}) is measured from the work plane to the floor.

To apply the zonal cavity method, it is necessary to determine a parameter known as the "cavity ratio" for each of the three cavities. Following is the formula for determining the cavity ratio:

$$\text{cavity ratio} = \frac{5\,\text{h(room length} + \text{room width)}}{\text{room length} \times \text{room width}}$$

where h = h_{cc} for ceiling cavity ratio (CCR)
h = h_{rc} for room cavity ratio (RCR)
h = h_{fc} for floor cavity ratio (FRC)

4.2.2 *Detailed lumen method*

Because of the ease of application of the lumen method which yields the average illumination in a room, it is usually employed for larger areas, where the illumination is substantially uniform. The lumen method is based on the definition of a footcandle equaling 1 lm/ft^2

$$\text{footcandle} = \frac{\text{lumen striking an area}}{\text{square feet of area}}$$

In order to take into consideration such factors as dirt on the luminaire, general depreciation in lumen output of the lamp, and so on, the above formula is modified as follows:

$$\text{footcandle} = \frac{\text{lamps/luminaire} \times \text{lumens/lp} \times \text{CU} \times \text{LLF}}{\text{area/luminaire}}$$

Table 4.4 Weighting Factors for Selecting Specific Illuminance Within Ranges D through I

Task or worker characteristics	Weighting factor		
	–1	0	≠1
Worker's age (average)	Under 40	40–55	Over 55
Speed or accuracy[a]	Not important	Important	Critical
Reflectance of task background	> 70%	30–70%	< 30%

Source: *IES Lighting Handbook*, 6th ed.

Note: Weighting factors are based on worker and task information.

1. Assign the appropriate factor for each characteristic.
2. Add the three weighting factors and refer to Table 4.2, Categories D through I:
   a. If the algebraic sum is –2 or –3, use the lowest range value.
   b. If the algebraic sum is –1, 0 or +1, use the middle range value
   c. If the algebraic sum is +2 or +3, use the highest range value.

a. Evaluation of speed and accuracy requires that time limitations, the effect of error on safety, quality, cost, etc., be considered. For example, leisure reading imposes no restrictions on time, and errors are seldom costly or unsafe. Reading engineering drawings or a micrometer requires accuracy and, sometimes, speed. Properly positioning material in a press or mill can impose demands on safety, accuracy, and time.

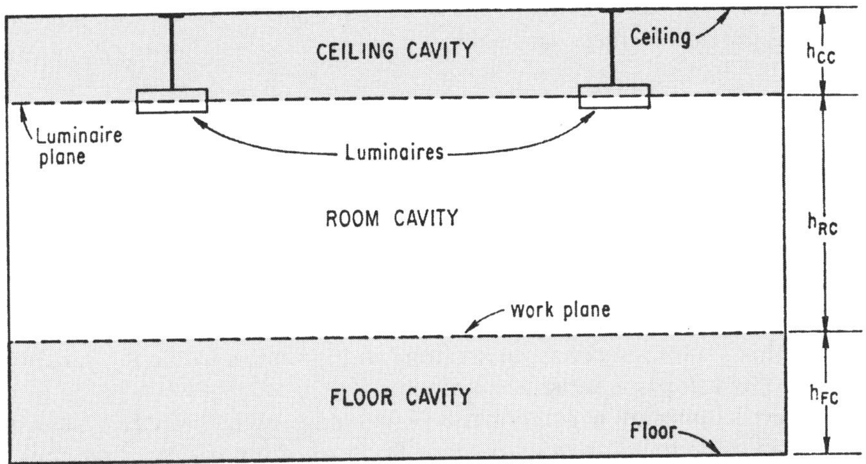

Figure 4.1 Basic cavity divisions of space.

In using the detailed lumen method, the following key steps should be taken:

1. Determine the required level of illuminance.
2. Determine the CU (coefficient of utilization), which is the ratio of the lumens reaching the working plane to the total lumens generated by the lamps. There is a factor that takes into account the efficiency and the distribution of the luminaire, its mounting height, the room proportions, and the reflectances of the walls, ceiling, and floor. Rooms are classified according to shape by ten room cavity numbers. The cavity ratio can be calculated using the formula given in Section 4.2.1. The CU is selected from tables prepared for various luminaires by manufacturers.
3. Determine the LLF (light loss factors). The final light loss factor is the product of all the contributing loss factors. Lamp manufacturers rate filament lamps in accordance with their output when the lamp is new; vapor discharge lamps (fluorescent mercury, and other types) are rated in accordance with their output after 100 h of burning.
4. Calculate the number of lamps and luminaires required:

$$\text{no. of lamps} = \frac{\text{footcandles} \times \text{area}}{\text{lumnens/lp} \times \text{CU} \times \text{LLF}}$$

$$\text{no. of luminaires} = \frac{\text{no. of lamps}}{\text{lamps/luminaire}}$$

5. Determine the location of the luminaires. Luminaire locations depend on the general architecture, size of bays, type of luminaires, position of previous outlets, and so on.

4.2.3 Point-by-point method

Although, currently, lighting computations emphasize the zonal cavity method, there is still considerable merit in the point-by-point method. This method lends itself especially well to calculating the illumination level at a particular point where total illumination is the sum of general overhead lighting and supplementary lighting. In this method, information from luminaire candlepower distribution curves must be applied to the mathematical relationship. The total contribution from all luminaires to the illumination level on the task plane must be summed.

Direct illumination component — The angular coordinate system is most applicable to continuous rows of fluorescent luminaires. Two angles are involved: a longitudinal angle *a* and a lateral angle *b*. Angle *a* is the angle between a vertical line passing through the seeing task to the end of the rows of luminaires. Angle *a* is easily determined graphically from a chart showing angles *a* and *b* for various combinations of V and H. Angle *b* is the angle between the vertical plane of the row of luminaires and a tilted plane containing both the seeing task and the luminaire or row of luminaires. Figure 4.2 shows how angles *a* and *b* are defined. The direct illumination component for each luminaire or row of luminaires is determined by referring to the table of direct illumination components for the specific luminaire. The direct illumination components are based on the assumption that the luminaire is mounted 6 ft above the seeing task. If this mounting height is other than 6 ft, the direct illumination component shown in Table 4.5 must be multiplied by 6/V, where V is the mounting height above the task. Thus the total direct illumination component would be the product of 6/V and the sum of the individual direct illumination component of each row.

Reflected illumination components — On the horizontal surface: This is calculated in exactly the same manner as the average illumination using the lumen method, except that the RRC (reflected radiation coefficient) is substituted for the coefficient of utilization.

$$FC_{RH} = \frac{\text{lamps/luminaire} \times \text{lumens/lp} \times \text{RRC} \times \text{LLF}}{\text{area/luminaire}}$$

where

$$\text{RRC} = LC_W + \text{RPM}(LC_{CC} - LC_W)$$

where LC_W = wall luminance coefficient
LC_{CC} = ceiling cavity luminance coefficient
RPM = room position multiplier

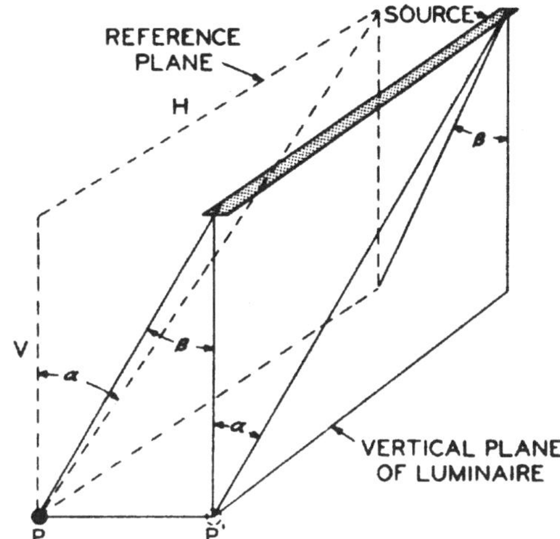

Figure 4.2 Definition of angular coordinate system for direct illumination component.

The wall luminance coefficient and the ceiling cavity luminance coefficient are selected for the appropriate room cavity ratio and proper wall and ceiling cavity reflectances from the table of luminance coefficient in the same manner as the coefficient of utilization. The room position multiplier is a function of the room cavity ratio and of the location in the room of the point where the illumination is desired. Table 4.6 lists the value of the RPM for each possible location of the part in the rooms of all room cavity ratios.

Figure 4.3 shows a grid diagram that illustrates the method of designating the location in the room by a letter and a number.

On the vertical surfaces: To determine illumination reflected to vertical surfaces, the approximate average value is determined using the same general formula, but substituting WRRC (wall-reflected radiation coefficient) for the coefficient of utilization:

$$FC_{RV} = \frac{\text{lamps/luminaire} \times \text{lumens/lp} \times \text{WRRC} \times \text{LLF}}{\text{area/luminaire(on work plane)}}$$

where

$$\text{WRRC} = \frac{\text{wall luminance coefficient}}{\text{average wall reflectance}} - \text{WDRC}$$

where WDRC is the wall direct radiation coefficient which is published for each room cavity ratio together with a table of wall luminance coefficients (see Table 4.5 for a specific type of luminance).

Table 4.5 Direct Illumination Components for Category III Luminaire (Based on F40 Lamps Producing 3100 lm)

Direct illumination components

β	5	15	25	35	45	55	65	75	5	15	25	35	45	55	65	75
α	Vertical surface illumination footcandles at a point on a plane parallel to luminaires								Vertical surface illumination footcandles at a point on a plane perpendicular to luminaires							
0-10	0.9	2.6	3.6	3.9	3.3	1.9	0.7	0.1	0.9	0.8	0.7	0.5	0.3	0.1	—	—
0-20	1.8	5.0	7.0	7.7	6.6	3.8	1.5	0.2	3.6	3.2.	2.7	1.9	1.2	0.5	0.1	—
0-30	2.6	7.2	10.1	11.3	9.8	5.7	2.3	0.3	7.7	7.0	5.8	4.3	2.7	1.1	0.3	—
0-40	3.2	9.0	12.8	14.5	12.9	7.7	3.2	0.5	12.6	11.6	9.7	7.5	4.9	2.1	0.6	—
0-50	3.7	10.3	14.9	17.1	15.7	9.6	4.3	0.7	17.8	16.6	14.2	11.2	7.7	3.4	1.1	0.1
0-60	4.0	11.2	16.3	18.8	17.6	11.3	5.5	1.0	22.6	21.2	18.4	14.7	10.4	5.1	1.9	0.2
0-70	4.1	11.6	17.0	19.8	18.9	12.7	6.8	1.4	26.2	24.7	21.8	17.8	13.1	7.2	3.2	0.3
0-80	4.1	11.7	17.3	20.2	19.4	13.3	7.4	1.9	28.2	26.7	23.8	19.7	14.9	8.7	4.3	0.8
0-90	4.1	11.7	17.3	20.2	19.4	13.4	7.5	2.0	28.6	27.1	24.2	20.1	15.3	9.1	4.7	1.1

Footcandles at point on workplane

α	5	15	25	35	45	55	65	75
0-10	10.6	9.5	7.6	5.5	3.3	1.3	0.3	—
0-20	20.6	18.5	14.9	10.9	6.6	2.6	0.7	—
0-30	29.4	26.5	21.6	16.0	9.8	4.0	1.1	—
0-40	36.5	33.1	27.4	20.6	12.9	5.4	1.5	—
0-50	41.8	38.1	31.9	24.3	15.7	6.7	2.0	0.1
0-60	45.2	41.3	34.8	26.8	17.6	7.9	2.6	0.2
0-70	46.9	43.0	36.4	28.3	18.9	8.9	3.2	0.3
0-80	47.4	43.6	36.9	28.8	19.4	9.3	3.5	0.4
0-90	47.5	43.7	37.0	28.8	19.4	9.3	3.5	0.4

Category III

2 T-12 lamps — any loading
For T-10 lamps — C.U. × 1.02

Luminance coefficients for 20% effective floor cavity reflectance

Ceiling Cavity		Reflectances											
		80		50		10		80		50		10	
Walls		50	30	50	30	50	30	50	30	50	30	50	30
WDRC	RCR	Wall luminance coefficients						Ceiling cavity luminance coefficients					
.281	1	.246	.140	.220	.126	.190	.109	.230	.209	.135	.124	.025	.023
.266	2	.232	.127	.209	.115	.182	.102	.222	.190	.130	.113	.024	.021
.245	3	.216	.115	.196	.105	.172	.095	.215	.176	.127	.105	.024	.020
.226	4	.202	.102	.183	.097	.161	.088	.209	.164	.124	.099	.023	.019
.212	5	.191	.097	.173	.090	.154	.082	.204	.156	.121	.094	.023	.018
.196	6	.178	.090	.163	.084	.145	.076	.200	.149	.118	.090	.022	.017
.182	7	.168	.083	.153	.078	.136	.071	.194	.144	.115	.087	.022	.017
.170	8	.158	.077	.145	.072	.130	.066	.190	.139	.113	.085	.021	.016
.159	9	.150	.072	.138	.068	.123	.062	.185	.135	.110	.082	.021	.016
.149	10	.141	.068	.130	.064	.116	.059	.180	.131	.107	.080	.020	.016

Table 4.6 Room Position Multipliers

	A	B	C	D	E	F		A	B	C	D	E	F
	Room cavity ratio = 1							**Room cavity ration = 6**					
0	.24	.42	.47	.48	.44	.48	0	.20	.23	.26	.28	.29	.30
1	.42	.74	.81	.83	.84	.84	1	.23	.26	.29	.31	.33	.36
2	.47	.81	.90	.92	.93	.93	2	.26	.29	.35	.37	.38	.40
3	.48	.83	.92	.94	.95	.95	3	.28	.31	.37	.39	.41	.43
4	.48	.84	.93	.95	.96	.97	4	.29	.33	.38	.41	.43	.45
5	.48	.84	.93	.95	.97	.97	5	.30	.36	.40	.43	.45	.47
	Room cavity ratio = 2							**Room cavity ratio = 7**					
0	.24	.36	.42	.44	.46	.46	0	.18	.21	.23	.25	.26	.27
1	.36	.51	.60	.63	.66	.68	1	.21	.23	.26	.28	.29	.30
2	.42	.60	.68	.72	.78	.83	2	.23	.26	.30	.32	.33	.34
3	.44	.63	.72	.77	.82	.85	3	.25	.28	.32	.34	.35	.36
4	.46	.66	.78	.82	.85	.86	4	.26	.29	.33	.35	.37	.37
5	.46	.68	.83	.85	.86	.87	5	.27	.30	.34	.36	.37	.38
	Room cavity ratio = 3							**Room cavity ratio = 8**					
0	.23	.32	.37	.40	.42	.42	0	.17	.18	.21	.22	.22	.23
1	.32	.40	.48	.51	.53	.57	1	.18	.20	.23	.25	.26	.26
2	.37	.48	.58	.61	.64	.67	2	.21	.23	.26	.27	.28	.29
3	.40	.51	.61	.65	.69	.71	3	.22	.25	.27	.29	.30	.30
4	.42	.53	.64	.69	.73	.75	4	.22	.26	.28	.30	.31	.32
5	.42	.57	.67	.71	.75	.77	5	.23	.26	.29	.30	.31	.32
	Room cavity ratio = 4							**Room cavity ratio = 9**					
0	.22	.28	.32	.35	.37	.37	0	.15	.17	.18	.19	.20	.20
1	.28	.33	.40	.42	.44	.48	1	.17	.18	.20	.21	.22	.23
2	.32	.40	.48	.50	.52	.57	2	.18	.20	.23	.24	.25	.25
3	.35	.42	.50	.54	.58	.61	3	.19	.21	.24	.25	.26	.26
4	.37	.44	.52	.58	.62	.64	4	.20	.22	.25	.26	.26	.27
5	.37	.48	.57	.61	.64	.66	5	.20	.23	.25	.26	.27	.27
	Room cavity ratio = 5							**Room cavity ratio = 10**					
0	.21	.25	.28	.31	.33	.33	0	.14	.16	.16	.17	.18	.18
1	.25	.29	.33	.36	.38	.42	1	.16	.17	.18	.19	.19	.20
2	.28	.33	.40	.42	.44	.48	2	.16	.18	.19	.21	.22	.22
3	.31	.36	.42	.46	.49	.52	3	.17	.19	.21	.22	.23	.23
4	.33	.38	.44	.49	.52	.54	4	.18	.19	.22	.23	.23	.24
5	.33	.42	.48	.52	.54	.56	5	.18	.20	.22	.23	.24	.25

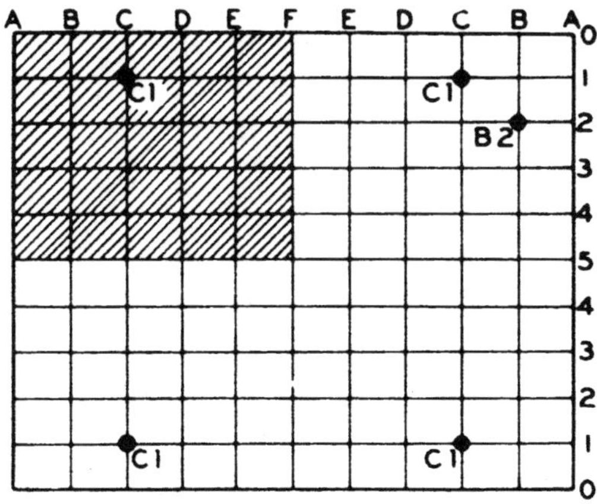

Figure 4.3 Grid diagram for locating points on the work plane.

4.2.4 *Typical examples*

Using the lumen method — The problem is given briefly as follows: to determine the number of fixtures which would be required to illuminate properly the hexagonal (six-sided) lobby, and that each wall of this space is 20 ft long. The ceiling is white acoustical tile, while the walls are dark paneling, with matching doors. The floor is of terrazzo tile. (Figure 4.4 shows

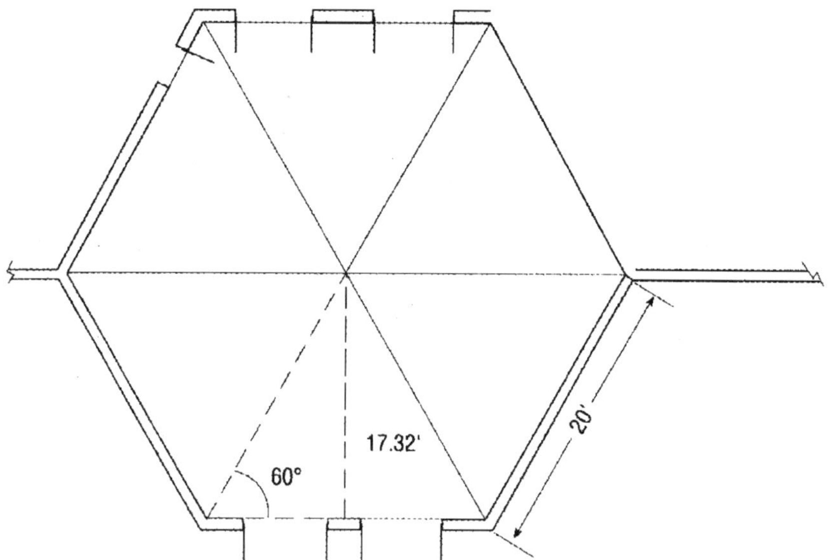

Figure 4.4 A hexagonal vestibule.

the geometry of the lobby.) For reasons of economy, it is suggested to use a fluorescent down-light utilizing the relatively new 13-W twin tube compact fluorescent lamp. Figure 4.5 shows a cross section of a typical fixture. The lamp produces 900 lm, according to the manufacturer's data.

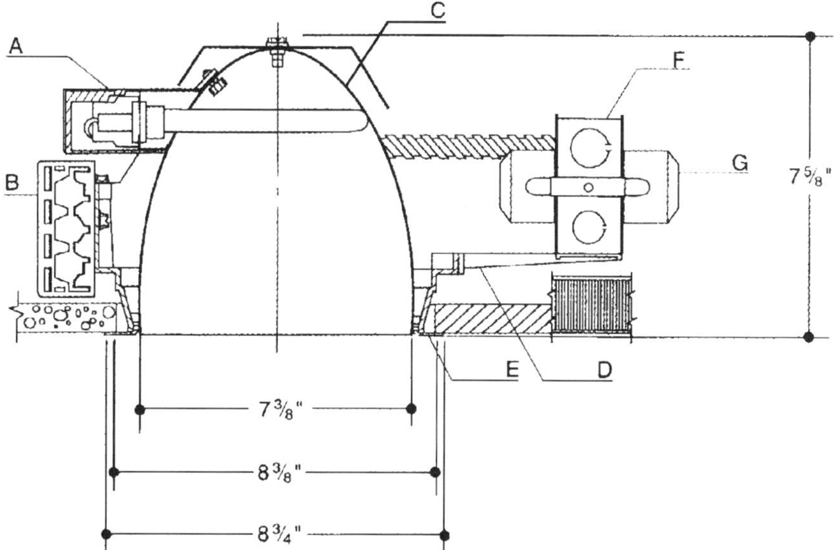

Figure 4.5 Typical fluorescent downlight. (A) **Socket**: heat-dissipating die cast aluminum socket for two 13-W compact fluorescent lamps. (B) **Universal mounting bracket**: exclusive universal mounting bracket; adjusts 5 in. vertically from inside or outside of fixture; bar hangers permit horizontal adjustment. (C) **Reflector**: 050 spun aluminum clear or gold specular Alzak finish. (D) **Housing/mounting frame**: one-piece precision die cast aluminum 1$^1/_2$ in. deep collar accommodates varying dimensions of ceiling materials. (E) **Trim ring**: high-impact polymer; stain white finish. (F) **Junction box**: U.L. listed for through-branch circuit wiring; positioned to allow straight conduit runs; five $^1/_2$-in. and two $^3/_4$-in. knockouts provided. (G) **Ballast**: type 1 potted ballast; one per lamp. (Courtesy of Halo Lighting.)

The design process starts by determining the room RCR. From Figure 4.4, the area of each of the six equilateral triangles is 20 × Sin 60° × 20 = 346.4 ft^2. Therefore the area of this room is 346.4 × 6 = 2078.4 ft^2.

The cavity ratio formula under Section 4.2.1 can also be written as follows:

$$\text{RCR} = \frac{2.5 \times \text{H} \times \text{room perimeter}}{\text{room area}}$$

We assume that we want to take the footcandle reading at the floor so the room cavity height is 12 ft (the height of the ceiling). Using the above formula,

$$\text{RCR} = \frac{2.5 \times 12 \times 6 \times 20}{2078.4} = 1.7$$

Next item is to determine the fixture coefficient of utilization. Since the ceiling is white, its reflectance value would be 80%. The wall is dark, so its reflectance is perhaps 10%; and the floor reflectance would be about 20% for terrazzo type.

To evaluate the LDD (luminaire dirt depreciation), assume the lobby to be a fairly clean area, and therefore LDD from the *IES Handbook* yields a value of 0.80. The other factor is LLD (lamp lumen depreciation) which can be found in the lamp manufacturer's data sheet. For a lobby, a lighting level of 10 fc would be appropriate.

$$\text{No. of luminaires} = \frac{10 \times 2078.4}{2 \times 900 \times 0.64 \times 0.86 \times 0.80} = 26$$

Having determined the minimum number of luminaires required to produce 10 fc, we should proceed to work out a symmetrical arrangement of fixtures for the space. It is further important to apply high power factor ballast for the 13-W lamp to reduce the current drawn and, consequently, the energy savings from the system.

Using point-by-point method — Calculate the illumination at a point assuming that four rows of six 4-ft luminaires (for which data are shown in Table 4.5) are surface mounted on 8-ft centers in a room 28 × 30 ft. Assume that the ceiling reflectance (and also that of the ceiling cavity since the luminaires are ceiling mounted) is 80% and that of the wall is 50%. Floor cavity reflectance is 20%. The mounting height of the luminaires is 8 1/2 ft above the work plane. The initial illumination on the horizontal work plane at point P is desired (see Figure 4.6, a typical luminaire layout plan, for this example).

Calculation of direct component: First let us determine angle *a* (for both ends of the rows of luminaires) and angle *b*. For angle *a*, H is 10 ft for *a1* and 14 ft for *a2*. The vertical distance V is 8 1/2 ft. For angle *b*, H is 12 ft for rows A and D and 4 ft for rows B and C. The vertical distance is still 8 1/2 ft. Refer to Table 4.5 for data on the direct illumination component. Table 4.7 summarizes the results of various components as found from the data in Table 4.5. Since the direct illumination component table is all based on a mounting height of 6 ft above the point, and in this case the luminaires are actually 8 1/2 ft above point P, it is necessary to multiply the total footcandles by 6/8.5. The resultant direct component is 114.8 fc.

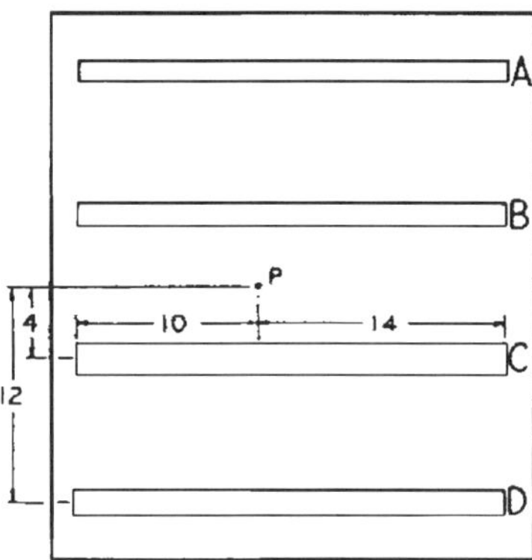

Figure 4.6 A typical luminaire layout plan.

Calculation of reflected component: The RCR for this room is 3.0 and the area per luminaire is 35 ft². Using the formula for computing the initial value of the reflected illumination component on the horizontal, FC is

$$FC_{RH} = 2 \times 3200 \times \text{reflected radiation coefficient}/35$$

The reflected radiation coefficient

$$\begin{aligned} RRC &= LC_W + RPM(LC_{CC} - LC_W) \\ &= 0.216 + 0.75(0.215 - 0.216) \\ &= 0.215 \end{aligned}$$

RPM is taken from Table 4.6 at point E5, and $FC_{RH} = 39.3$.
The total illumination at point P is 114.8 + 39.3 = 154.1 fc.

Table 4.7 Summary of Direct Illumination Components

Row	α_1	α_2	β	Direct illumination component: Front left end	Front right end	Total
A	50	60[a]	55	6.7	7.9	14.6
B	50	60	25	31.9	34.8	66.7
C	50	60	25	31.9	34.8	66.7
D	50	60	55	6.7	7.9	14.6
						162.6

a. Actually α_2 is 59 but is rounded off to 60.

4.3 Computer programs for illumination design

4.3.1 Guidelines for choosing features for zonal cavity and point-by-point design software

There are some desirable features of illumination software that are general in nature and common to both zonal cavity and point-by-point programs. Illuminating software should be easy to use and understand. The easiest software to use is generally menu driven with a full screen, "fill-in-the-blank" style of data entry. Good illumination software should also "dynamically" check data as it is entered so that mistakes are minimized. Ideally, the software should have some provision for storing input data so that it is easy to make revisions as a project changes. Finally, the reports printed should be nicely formatted and easily understandable.

Evaluating zonal cavity software — All illumination design programs that use the zonal cavity method developed by the Illuminating Engineering Society (IES) are commonly referred to as zonal cavity-based programs. The basic function of a zonal cavity program is to calculate the number of luminaires required for a desired average level of illuminance in footcandles. In accomplishing this goal, a zonal cavity program should provide a library of luminaire data that can be easily updated and maintained by the user. It is ideal if the luminaire data file is not restricted to just one brand of luminaires.

One advanced feature a lighting engineer should look for in a zonal cavity program is the ability to store data for all the rooms in a project. Other advanced features to look for include the ability to perform economic comparisons of luminaires, customized luminaire schedule reports, bill of materials with both labor and material costs, and graphic luminaire layout reports. Luminaire layout reports should take into account luminaire spacing-to-mounting height ratios as well as special design requirements such as end-to-end spacing.

The basic formula for calculating the number of luminaires required for a desired level of illuminance is given in Section 4.2.2.

Specifying CU values: The above formula is so simple to use with the exception of the CU values. The specific CU value used for a given calculation depends on many factors, including both room and luminaire parameters. The process for determining the correct CU to use is so tedious that if a program does not do this automatically, it is not performing much of a service for the engineer. Fortunately, most zonal cavity programs do automatically determine the proper CU value to use. These programs must maintain large photometric data files that contain CU values for numerous luminaires. However, determining the proper CU value is more than just a table lookup procedure. It involves interpolation, sometimes up to three ways for various room cavity ratios, wall reflectances, and ceiling reflectances. In addition, CU values often have to be adjusted for effective floor cavity reflectances other than 20%. It is therefore important to make sure that the zonal cavity program chosen for use does interpolate proper CU values.

Evaluating point-by-point programs — Once zonal cavity calculations are automated, the illuminating engineers can consider using a point-by-point calculation program. Many of the same considerations given for zonal cavity software also apply to point-by-point software. This is especially true concerning the feature of reading IES formatted photometric data files. However, due to the nature of point-by-point calculations, there are many extra factors to consider.

Whereas the purpose of zonal cavity software is to estimate the number of luminaires required for an average level of illuminance, point-by-point software calculates illuminance levels at specified points in the lighted area for a given luminaire scenario. Unlike zonal cavity software, point-by-point programs require the engineer to specify how many luminaires to evaluate. Because of this, point-by-point software is primarily intended to perform a "what if" analysis for a given luminaire layout. The idea is that the engineer "grids" his lighting area by deciding on a uniform spacing of points that cover the entire area. He then conceives a luminaire layout and tests that layout using the point-by-point software, which reports back with detailed horizontal and vertical footcandles at every point defined in the lighted area. These reports allow the engineer to know exactly how the lighting system will perform. If not entirely satisfied, the engineer can make adjustments and calculate again. This process continues until an optimal design is reached.

Comparison of two programs — Illumination using point-by-point software is much more precise than using zonal cavity software. It is also more versatile since it can be used for both indoor and outdoor applications, while zonal cavity is only appropriate for indoor design. Zonal cavity calculations are simple and fast, but point-by-point calculations are complex and slow to perform, even by computer. It is important to have fast-calculating software, especially for projects that may require numerous runs for optimization purposes.

However, there are other desirable features in a point-by-point program. One area of particular concern is data capacity. The better program can allow areas to be gridded with at least 1600 points. One option that helps to extend the size of the area that can be gridded is the concept of "masking". Many programs allow the user to define a rectangular area with some sections masked or not gridded with points. Some programs even allow for easy gridding of nonrectangular areas such as circles, triangles, and other geometric shapes.

Capable point-by-point software should allow for hundreds of luminaires illuminating a lighted area. The engineer should also have the capability of specifying exact positioning data such as height above ground, rotation, tilt, and roll for each luminaire.

4.3.2 Indoor vs. outdoor programs

Due to additional complexities of handling reflective surfaces, some point-by-point programs cannot be used for indoor calculations. When reviewing descriptions of programs, it is important to note this. If a program can handle indoor applications, there is usually provision for the point of reflectance values for walls, ceilings, and floors.

Although printing out horizontal and vertical footcandle levels at each point is the main output of a point-by-point program, other output is sometimes also given. For example, the better point-by-point programs can calculate equivalent sphere illumination (ESI), illumination with body shadow, and visual comfort probability (VCP) levels. Some programs even have graphic reports that show footcandle levels in shades of gray. Still others can produce plan view and perspective drawings of the lighted area and luminaire layout. A few programs have the ability to work with popular computer-aided drafting (CAD). The best zonal cavity programs will store multiple room data for each project, read IES formatted luminaire files, print a luminaire schedule report, perform economic comparisons relating initial cost and energy consumption, print a bill of materials with cost estimates, print a graphic luminaire layout report, and quickly calculate luminaire requirements.

Any illumination design involving outdoor area automatically requires a point-by-point program. The zonal cavity methods do not apply for outdoor applications. If an engineer encounters unusual elements in his indoor applications such as indirect lighting and large open areas, a point-by-point program is necessary.

4.3.3 Computation of ESI values

Despite the fact that the point-by-point program allows for a particular purpose, this design approach still restricts the illuminating engineers to determining "raw" footcandles and does not provide a measure of whether satisfactory visibility will be produced. With typical reading and writing tasks, visual performance loss caused by veiling reflections should be taken into account. This can only be achieved by ESI computations. The ESI report provides the mathematical details for computing ESI from candlepower data which are the same as those for the point-by-point computations. However, the viewing direction of the observer must be specified, as ESI can change substantially with the viewer's orientation. ESI computation is similar to point-by-point computation except the direction of all light rays is analyzed and associated with the reflecting characteristics of the task.

The fundamental qualities involved in the computation of ESI due to a single luminaire are expressed in a form that permits separation of the two variables that give the luminaire's position with respect to an observer. The total effect of all the luminaires in the layout is expressed in a single easily evaluated equation, which is a function of the two variables that give the observer's position. Specification of luminaire placement and orientation is achieved by a simple data input technique. For each of the four viewing directions, the maximum, minimum, average, and mean deviation of ESI values are calculated. These quantities are also calculated for all four directions taken together. The grid point may be as large as 20 × 20, which will give 400 points. This will yield a total of 1600 values of ESI. All ESI are printed in an array that corresponds to their positions in the plan view of the room. The values of background luminance (LB), contrast rendition factor (CRF), lighting effectiveness factor (LEF), and effective visibility level are printed in the same format. Figure 4.7 shows a typical room with six points marked, and Table 4.8 gives the resultant sample computer printout.

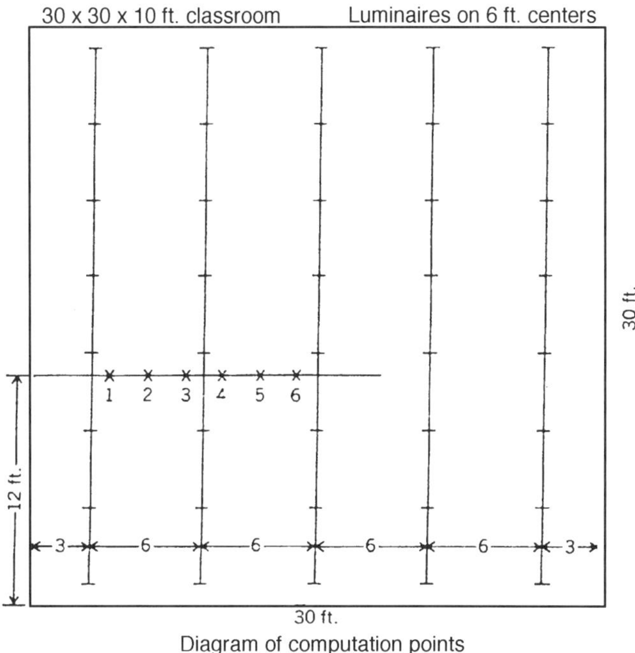

Figure 4.7 Typical 30 × 30 × 10 room with six points as marked.

4.3.4 *VCP values*

Many factors are involved in evaluation of the relative comfort of a lighting installation: shape and size of room; reflectances of room surfaces; illuminance level; type, size, and light distribution of luminaire used; number and location of luminaires; luminaires and their relationship in the entire field of view; location and line of sight of observer; and differences in observer sensitivity to glare. A comprehensive standard evaluation procedure taking all of the foregoing factors into account has come about as the result of numerous extensive investigations. This procedure provides a VCP rating of a given system of illumination. The rating is in terms of the percentage of people who will be expected to find the given illumination system acceptable when they are seated in the most undesirable location.

Table 4.8 Computer Printout on ESI and Related Computations

	Location X	Location Y	Orientation angle	FC[a]	FC	LB	LT	CRF	ESI	LEF
1	4.00	12.00	0.0	128.87	116.26	101.28	86.54	.869	40.02	.344
			90.0		114.95	96.17	81.35	.920	61.15	.532
2	6.00	12.00	0.0	139.02	135.63	116.16	97.89	.939	82.02	.605
			90.0		122.48	105.09	89.56	.882	46.61	.381
3	8.00	12.00	0.0	145.06	132.43	114.75	97.63	.890	53.49	.404
			90.0		134.84	115.13	97.37	.921	70.83	.525
4	10.00	12.00	0.0	147.29	134.71	116.61	99.17	.893	56.05	.414
			90.0		133.86	114.68	97.19	.911	64.88	.485
5	12.00	12.00	0.0	147.44	144.18	123.15	103.61	.947	92.16	.639
			90.0		131.47	113.89	96.95	.888	52.03	.396
6	14.00	12.00	0.0	148.68	136.10	117.75	100.09	.895	57.84	.425
			90.0		139.29	119.49	100.99	.924	74.99	.538

a. FC (this column) is the footcandle level for no body shadow. All other values include body shadow. LB and LT are the luminances of the background and task, respectively.

By means of several procedures outlined in the *IES Lighting Handbook* it is possible and useful to study proposed illumination system designs from the standpoint of VCP by preparing tables such as Table 4.9.

4.4 Definition of basic terms and factors in illumination

Illuminance — Illuminance is the density of luminous flux on a surface, expressed in either footcandles (lumens/ft^2) or lux (lx) (1 lux = 0.0929 fc).

Luminance (or photometric brightness) — Luminance is the luminous intensity of a surface in a given direction per unit of projected area

Table 4.9 Typical VCP Values

Wall reflectance: 50%
Ceiling cavity reflectance: 80%
Effective floor cavity reflectance: 20%

Work plane illumination: 100 fc
Luminaire no. 00
Room size 60 × 30 ft

Mounting height above the floor (ft)	Luminaires Lengthwise	Luminaires Crosswise
8.5	68	67
10	69	69
13	71	70
16	74	73

of the surfaces, expressed in candelas per unit area or in lumens per unit area.

Reflectance — Reflectance is the ratio of the light reflected from a surface to that incident upon it. Reflection may be of several types, the most common being specular, diffuse, spread, and mixed.

Glare — Glare is any brightness that causes discomfort, interference with vision, or eye fatigue.

Color rendering index (CRI) — In 1964 the CIE (Commission Internationale de l'Eclairage) officially adopted the IES procedure for rating lighting sources and developed the current standard by which light sources are rated for their color-rendering properties. The CRI is a numerical value of the color comparison of one light source to that of a reference light source.

Color preference index (CPI) — The CPI is determined by a similar procedure to that used for the CRI. The difference is that CPI recognizes the very real human ingredient of preference. This index is based on people's preference for the coloration of certain identifiable objects, such as complexions, meat, vegetables, fruits, and foliage, to be slightly different than their colors are in daylight. CPI indicates how a source will render colors with respect to how we best appreciate and remember those colors.

Luminous efficacy — The efficiency of light sources, derived by dividing the light output (in lumens) by the power input (in watts), is commonly denoted in lumens/watt (LPW).

Ballast factor (BF) — The ratio of the lamp lumen output on a commercial ballast to the lamp's rated light output.

Ballast efficacy factor (BEF) — Ballast efficacy factor for a specific lamp/ballast combination is calculated by dividing the percent rated light output by the measured input power in watts. The percent rated light output is found by multiplying the ballast factor by 100%. BEF is meant to compare the performance of ballasts on a specific lamp and is not of particular value in evaluating efficiency.

Equivalent sphere illumination (ESI) — ESI is a means of determining how well a lighting system will provide task visibility in a given situation. ESI may be predicted for many points in a lighting system through the use of any of several available computer programs, or measured in an installation with any of several different types of meters.

Visual comfort probability (VCP) — Discomfort glare is most often produced by direct glare from luminances that are excessively bright. Discomfort glare can also be caused by reflected glare, which should not be confused with veiling reflections, which cause a reduction in visual performance rather than discomfort. VCP is based in terms of the percentage of people who will be expected to find the given lighting system acceptable when they are seated in the most undesirable location.

Coefficient of utilization (CU) — CU is the ratio of the lumens reaching the work plane (assumed to be a horizontal plane 30 in. above the floor) to the total lumens generated by the light source. This is a factor that takes into account the efficiency and distribution of the luminaire, its mounting height, the room proportions, and the reflectance of the wall, ceiling, and floor.

Light loss factor (LLF) — The final LLF is the product of all the contributing loss factors. It is the ratio of the illumination when it reaches its lowest level at the task just before corrective action is taken, to the initial level if none of the contributing loss factors were considered. There are eight contributing loss factors that require consideration:

1. Ballast performance (ballast factor)
2. Voltage to luminaires
3. Luminaire reflectance and transmittance changes
4. Lamp outages
5. Luminaire ambient temperature
6. Heat-exchange luminaires
7. Lamp lumen depreciation (LLD)
8. Luminaire dirt depreciation (LDD)

Lamp lumen depreciation factor (LLD) — This factor is used in illumination calculations to quantify the output of light source at 70% of their rated life as a percentage of their initial output.

4.5 Factors and remedies

Quality of illumination pertains to the distribution of luminaires in the visual environment. The term is used in a positive sense and implies that all luminaires contribute favorably to visual performance. However, glare, diffusion, reflection, uniformity, color, luminance, and luminance ratio all have a significant effect on visibility and the ability to see easily, accurately, and quickly. Industrial installations of poor quality are easily recognized as uncomfortable and possibly hazardous. Some of the factors are discussed in more details below:

Direct glare

When glare is caused by the source of lighting within the field of view, whether daylight or electric, it is defined as direct glare. To reduce direct glare, the following suggestions may be useful:

1. To decrease the brightness of light sources or lighting equipment, or both
2. To reduce the area of high luminance causing the glare condition
3. To increase the angle between the glare source and the line of vision
4. To increase the luminance of the area surrounding the glare source and against which it is seen

To reduce direct glare, luminaires should be mounted as far above the normal line of sight as possible and should be designed to limit both the luminance and the quality of light emitted in the 45 to 85° zone because such light may interfere with vision. This precaution includes the use of supplementary lighting equipment. There is such a wide divergence of tasks and environmental conditions that it may not be possible to recommend a degree of quality satisfactory to all needs. In production areas, luminaires within the normal field of view should be shielded to at least 25° from the horizontal, preferably to 45°.

Reflected glare

Reflected glare is caused by the reflection of high luminance light sources from shiny surfaces. In the manufacturing area, this may be a particularly serious problem where critical seeing is involved with highly polished sheet metal, vernier scales, and machined metal surfaces. There are several ways to minimize or eliminate reflected glare:

1. Use a light source of low luminance, consistent with the type of work in process and the surroundings.
2. If the luminance of the light source cannot be reduced to a desirable level, it may be possible to orient the work so that reflections are not directed in the normal line of vision.
3. Increasing the level of illumination by increasing the number of sources will reduce the effect of reflected glare by reducing the proportion of illumination provided on the task by sources located in positions causing reflections.
4. In special cases, it may be practical to reduce the specular reflection by changing the specular character of the offending surface.

Distribution, reflection, and shadows

Uniform horizontal illuminance (maximum and minimum not more than one sixth above or below the average level) is usually desirable for industrial interiors to permit flexible arrangements of operations and equipment, and to assure more uniform luminance in the entire area. Reflections of light sources in the task can be useful provided that the reflections do not create reflected glare. In the machining and inspection of small metal parts, reflections can indicate faults in contours, make scribe marks more visible, and so on.

Shadows from the general illumination systems can be desirable for accenting the depth and forms of various objects, but harsh shadows should be avoided. Shadows are softer and less pronounced when large diffusing luminaires are used or the object is illuminated from many sources. Clearly defined shadows are distinct aids in some specialized operations, such as engraving on polished surfaces, some types of bench layout work, or certain textile inspections. This type of shadow effect can best be obtained by supplementary directional lighting combined with ample diffused general illumination.

Luminance and luminance ratios

The ability to see details depends on the contrast between the detail and its background. The greater the contrast difference in luminance, the more readily the seeing task is performed. The eye functions most comfortably and efficiently when the luminance within the remainder of the environment is relatively uniform. In manufacturing, there are many areas where it is not practical to achieve the same luminance relationships as easily as in offices.

Table 4.10 is shown as a practical guide to recommended maximum luminance ratios for industrial areas. To achieve the recommended luminance relationships, it is necessary to select the reflectances of all the finishes of the room surfaces and equipment as well as control of the luminance distribution of lighting equipment. Table 4.11 lists the recommended reflectance values for industrial interiors and equipment. High-reflectance surfaces are desirable to provide the recommended luminance relationships and high utilization of light.

Color quality of light

In general, for seeing tasks in industrial areas, there appears to be no effect upon visual acuity by variation in color of light. However, where color

Table 4.10 Recommended Maximum Luminance Ratios for Industrial Areas

	Environmental classification[a]		
	A	B	C
Between tasks and adjacent darker surroundings	3 : 1	3 : 1	5 : 1
Between tasks and adjacent lighter surroundings	1 : 3	1 : 3	1 : 5
Between tasks and more remote darker surfaces	10 : 1	20 : 1	[b]
Between tasks and more remote lighter surfaces	1 : 10	1 : 20	[b]
Between luminaires (or windows, skylights, etc.) and surfaces adjacent to them	20 : 1	[b]	[b]
Anywhere within normal field of view	40 : 1	[b]	[b]

a. A—Interior areas where reflectances of entire space can be controlled in line with recommendations for optimum seeing conditions.

B—Areas where reflectances of immediate work area can be controlled, but control of remote surround is limited.

C—Areas (indoor and outdoor) where it is completely impractical to control reflectances and difficult to alter environmental conditions.

b. Luminance ratio control not practical.

Table 4.11 Recommended Reflectance Values for Industrial Interiors and Equipment

Surfaces	Reflectance[a] (%)
Ceiling	80–90
Walls	40–60
Desk and bench tops, machines, and equipment	25–45
Floors	Not less than 20

a. Reflectance should be maintained as near as practical to recommended values.

discrimination or color matching is a part of the work process, such as in the printing and textile industries, the color of light should be carefully selected. Color always has an effect on the appearance of the work space and on the complexions of people. The illuminating system and the decorative scheme should be properly coordinated.

Veiling reflections

Figure 4.8 shows that light would reflect into the eyes of the viewer from the "offending zone" and defines the zone of veiling reflection. Veiling reflection would diminish visibility, but the viewer would be unaware of it. The contrast rendition factor (CRF) can be applied as a measure of the amount of veiling reflection.

Lighting effectiveness factor (LEF)

An overall lighting system efficiency factor considers both the quality of light as reference to equivalent sphere illumination and the effects of veiling reflections. Light patterns such as "batwing" can help solve veiling reflection problems. Figure 4.9 shows the light distribution curve of a typical batwing luminaire.

4.6 *Daylighting*

The daylighting contribution should be carefully evaluated and should always be coordinated with a planned electric lighting system.

Several basic terms used in dealing with daylighting are mentioned below.

Fenestration — Fenestration has at least three useful purposes in an industrial building:

1. For the admission, control, and distribution of daylight
2. For a distant focus for the eyes, which relaxes the eye muscles
3. To eliminate the dissatisfaction many people experience in completely closed-in areas

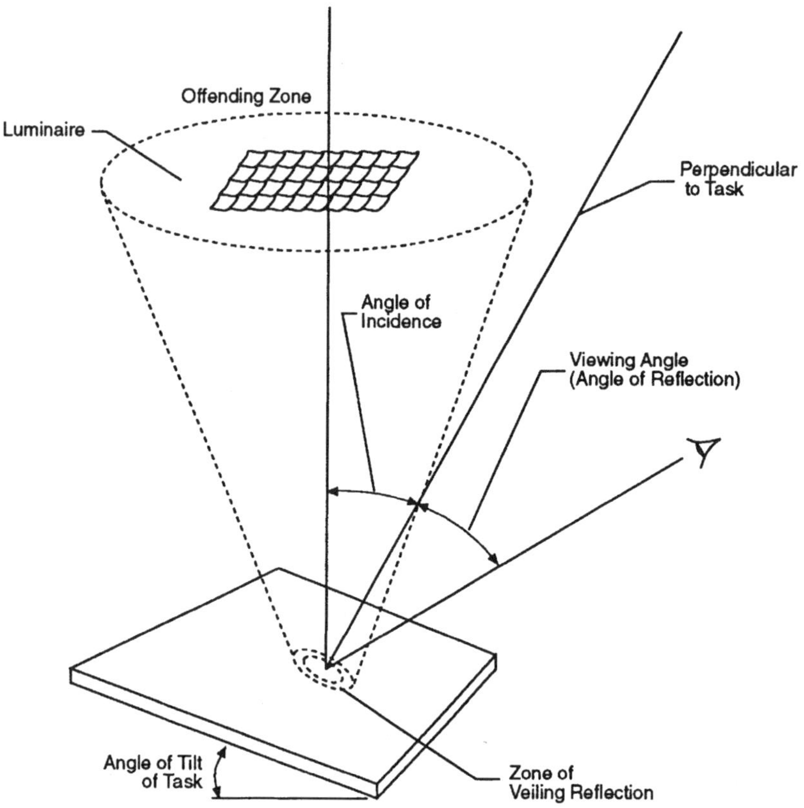

Figure 4.8 Diagram showing "offending zone" and zone of veiling reflection.

An adequate electric lighting system should always be provided because of the wide variation in daylight.

Building orientation — All fenestration should be equipped with a control device appropriate to any luminance problems. Special attention should be given to glare control for latitudes where fenestration frequently receives direct sunlight. Diffuse-glaring fixed or adjustable louvers are some of the control means that may be applied.

For an industrial building, windows in the sidewalls admit daylight and natural ventilation and afford occupants a view out. However, their uncontrolled luminance may be a problem. There are many control means to make the daylight useful to workers' seeing tasks, resulting in energy savings as the ultimate goal.

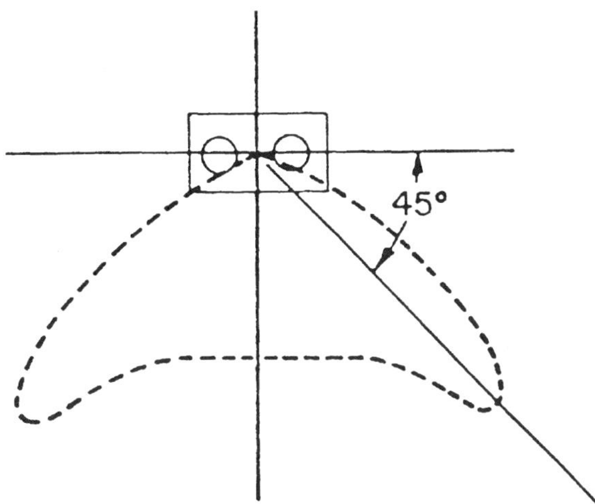

Figure 4.9 A typical "batwing" light distribution.

References

Allphin, W., Minimizing veiling reflections in lighting installations, *Plant Eng.*, June, 1972.

Chen, K., New concepts in interior lighting design, *IEEE Trans. Ind. App.*, pp.1179–1184, September/October, 1984.

Chen, K., *Industrial Power Distribution and Illuminating Systems*, Marcel Dekker, New York, 1990.

DiLaura, D.L., Whatever happened to equivalent sphere illumination, *Light. Des. and Appl.*, pp.17–18, November, 1982.

IES Computer Committee Report, Available lighting computer programs, *Light. Des. and Appl.*, September, 1986.

IES software survey, *Light. Des. and Appl.*, March, 1991.

Lighting Handbook, Westinghouse Electric Corporation, Bloomfield, NJ, May 1978.

Lighting Handbook, Application Volume, Illuminating Engineering Society, New York, 1987.

Mann, P., A new look at lighting calculations, *CEE News*, September, 1993.

Rowe, G.D., Determining illumination requirements, *Plant Eng.*, pp.69–72, February 4, 1982.

Sisson, W., Determining cavity ratios for zonal cavity lighting calculations, *Plant Eng.*, pp.68–69, November 27, 1970.

Smith, W., Evaluating lighting design software, *CEE News*, November, 1989.

chapter five

Daylighting and lighting controls

5.1 Daylighting

5.1.1 Introduction

Daylighting and controls create an ideal match to maximize peak electrical demand savings from illuminating systems operating at reduced power and associated reduction in chiller system power. The demand savings are more important than energy savings since daylighting provides energy reduction at the peak electricity usage time.

In addition there are many other benefits of daylighting. A connection with the outdoors is also important to our well-being. Sunlight interplays in buildings, adding variations and excitement to interior spaces, and it can enhance occupant comfort. It provides the best color rendering index and helps printing and machine shop works.

However, daylighting is not a total substitute for electric lighting. Contemporary lighting design practices require that illumination meeting a specified design criterion be provided whenever building is occupied. The sole use of daylighting does not meet such a criterion, because daylighting illumination levels change continually with time, season, and sky conditions. To exploit daylighting as a source of illumination, it is necessary to establish an interactive link between ambient lighting conditions and the electric lighting system. This can be achieved with a photoelectrically controlled lighting system based on the amount of prevailing daylight. Lighting control hardware that links electric lighting to available daylighting generally falls into two categories:

1. Continuously dimmable electric lighting systems controlled by photosensors, which continually adjust the electric lighting level in response to the amount of daylight striking the control photosensors. At the heart of the system is the controllable output ballast or dimming ballast.
2. Photo-relay-based systems, which automatically switch off perimeter lighting when daylighting is sufficient to meet lighting needs.

The lighting control system shown in Figure 5.7 can be used to achieve the above-described lighting level control scheme with the aid of photosensors located within the task zones. Photosensor information is used by the microcomputer to control zone lighting levels, which can be controlled from 100 to 0%. In zones where natural light is available, the system automatically takes advantage of this resource and incorporates it into the space. The electric lighting in daylight zones is used as fill-in lighting to provide even illumination across the space. The amount of usable daylight is dependent on the windows and skylight types and sizes, window and skylight treatments (drapes, blinds, glazing, etc.), building design, interior design, and furnishings.

5.1.2 *Daylighting techniques*

A number of techniques may be used to provide daylighting:

1. Sidelighting — The most familiar approach to daylighting technique is sidelighting. The quantity of light admitted from side openings, usually provided by windows or glass block, depends on the width and height of the opening above the working plane, the type of glazing, and any control elements, such as blinds or louvers. The effectiveness of sidelighting is limited to a distance into the room away from the windows of about 2.5 times the height of the opening (somewhat further with glass block). Direct sunlight is not preferred because it causes critical brightness and thermal problems, unless it is properly controlled by selective-transmitting materials, window orientation, or adjustable blinds or drapes. However, these types of control can limit the potential interior illumination at times when there is an overcast sky. In this case, supplementary lighting with daylight harvesting controls should be considered. Often the benefits of daylighting and winter heat gain through side windows outweigh the disadvantages of summer heat increase to be compensated for by air conditioning.
2. Toplighting — Skylights and clerestories provide about three times the amount of daylight as vertical windows, and since they can be placed closer to the center of an area, they provide more uniform lighting than windows can with sidelighting. Diffusing glazing materials, sometimes in combination with ceiling louvers or lenses, can be used to control the brightness of direct sunlight and high sky luminance to avoid reducing visual comfort. Total light transmission and toplight surface luminance can be balanced with glazing materials by selecting appropriate transmittance values. Sometimes skylights are applied improperly, causing excessive heat input into the building. However, recent studies show that skylights can be applied throughout the U.S. to provide daylight with annual heat transfer well within cost-effective considerations.

Hardware includes:

1. Light shelves — Light shelves are reflective horizontal surfaces that extend from the exterior to the interior of a building to increase the range of perimeter daylighting on the structure's south side and reduce the amount of heat that enters the space. Light shelves work best at high solar angles, but at lower angles the shelves need to extend further into the space to capture the sun's ray. Attempts in the 1980s to produce a variable-area light-reflecting assembly were not cost effective.
2. Prismatic glazing — It uses the principle of reflection to redirect the sun's ray to the ceiling or to exclude it from space altogether. Using a material with molded prisms, prismatic glazing works best at certain solar angles, but as the ray moves beyond the critical angles direct sunlight can cause glare. Since prismatic glazing obscures the view of the outdoors, it is usually placed at the top of a window.
3. Passive skylights — Passive skylights contain no moving parts and are a simple way to bring daylight into the upper floor of a deep building. Skylights can cause thermal problems and tend to lose light as a result of reflections in the shaft between the roof and the ceiling.
4. Sun pipe — This is an aluminum pipe lined with silver to reduce the losses from reflections. It uses a clear acrylic dome on the top and a translucent dome on the bottom to diffuse light into a space.

5.1.3 Designing for daylight

Introduction

In developing a preliminary building design, architects must first establish interior visual criteria and basic lighting performance requirements for their particular project. Next, they will have to determine the parameters of the available daylight for the particular location and select the appropriate daylight data to be used as a basis for design purposes. They are now ready to calculate daylight contributions for various design schemes.

Establishing a sky condition for design

Sky conditions are extremely variable for almost all localities. From morning to evening, from season to season, and from day to day, the amount of light available from the sun and sky is constantly changing. Additionally, very little measured data are available. It is impossible to establish a single set of conditions that can be used as an absolute design base. Generally the designer will want to deal with maximum and minimum sky conditions or with the condition which may be considered most prevalent.

Predicting interior daylighting

Interior illumination is determined for each of several conditions and added together for final results. For instance, illumination levels from windows and from skylights can be determined separately, and then added together for final results. Illumination from the sky and ground are determined separately and added for final results. More detailed discussions are given below:

1. Combination daylighting sources — Daylighting designs which combine side-wall fenestration in opposite walls, or designs which combine side-wall lighting and toplighting can be treated by superimposition. That is, calculate the workplane illumination from one source of daylight; calculate the illumination from the second source; and add the two results to get the total illumination on the workplane.
2. Combination daylight and electric — Similarly, building designs which combine daylighting and electric lighting can be treated by superimposition. The workplane illumination calculated for daylight can be added to that calculated for electric light.
3. Reflected ground light — Although the condition of the ground varies from side to side, thus a part of the designed environment, in these calculations light reflected from the ground, is considered as another light source. The light from the ground and from the sky must be separately determined.
4. Other prediction techniques — The daylight prediction technique described in the above is based on the lumen method commonly used for calculating the illumination from electric lighting and should be familiar to illuminating engineers. However, there are a number of other techniques for predicting interior daylighting, including graphic, mathematical, and simulation processes.

5.1.4 *Daylighting economics*

A well-designed daylighting application may realize annual energy cost savings of 20 to 30% compared to buildings without daylight design or controls. Electric energy used for lighting systems can be reduced as much as 70% during peak natural light period.

Predicting the economic effects of daylighting systems is difficult because of the many assumptions involved. In general, a sample calculation shows that energy costs saved per year can be approximately $0.25 per square foot of daylighted floor area.

5.1.5 *Practical method of utilizing daylight*

A practical method of utilizing daylighting to conserve energy is shown in Figure 5.1. Indicated herein is a method of wiring two luminaires with a two-lamp ballast to achieve three-step control. As more daylighting is available, more rows near the window are switched off. They may appear to be a crude method, but it is an economical approach that requires no elaborate control hardware. The wiring method responds to daylighting in the following manner:

Sequence	Illumination (%)
Power on to ballasts X and Y	100
Switch off power to either X or Y	50
Switch off power to both X and Y	0

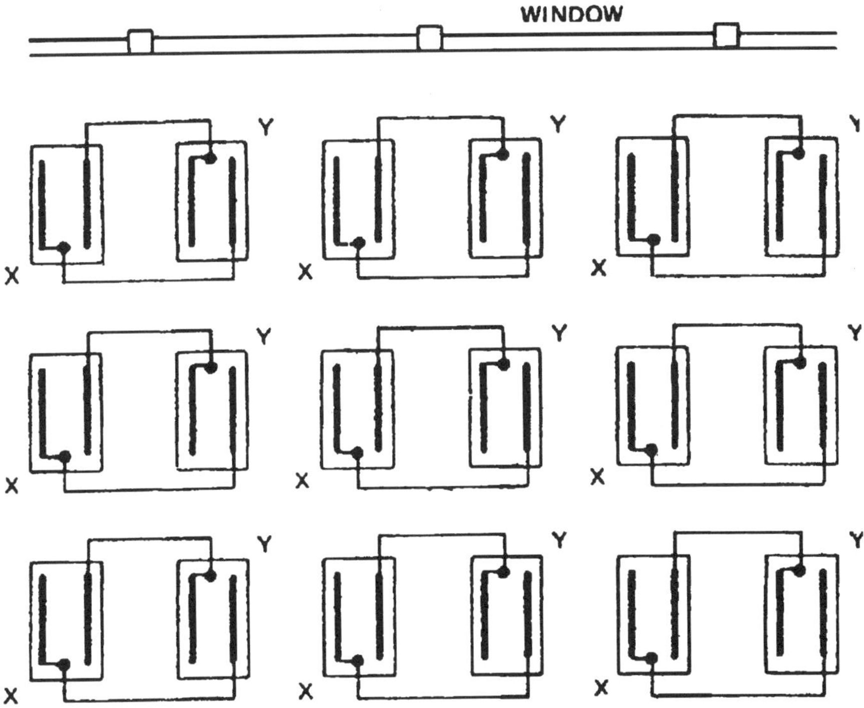

Figure 5.1 Method of wiring for three-step control. One two-lamp ballast controls one lamp in the luminaire and one lamp in the adjacent luminaire. Illumination of each row can be adjusted in 3-steps in response to daylighting in the following manner:

Sequence	Illumination (%)
Power on to ballasts X and Y	100
Switch off power to either ballast X or Y	50
Switch off power to both ballasts X and Y	0

As more daylighting is available more rows near the window are switched off.

5.2 Lighting controls

5.2.1 Introduction

Lighting controls have become increasingly sophisticated in recent years, mainly due to the dire need for energy savings and advancements in the solid-state control devices. As the cost of energy has continued to rise, increasing effort has gone into minimizing the energy consumption of lighting installations. This effort has evolved along three major directions: the development of new energy-efficient lighting equipment, the utilization of improved lighting design practice, and, finally, improvements in lighting control systems. The lighting industry has been introducing more efficient lighting components and systems, which cost more to install but result in a lower total cost (operating plus initial). End users begin to base their decisions on the ROI or the life-cycle cost. These decisions should lead to much greater application of cost-effective lighting control systems.

In the last decade, many new lighting control hardwares have appeared on the market. In this chapter we review all important types of lighting controls for illuminating systems and offer some guidelines for selection to suit individual needs and achieve optimum energy saving.

5.2.2 Types of controls

All lighting controls, whether a simple switch or a sophisticated programmable controller, can normally be classified in two basic categories: (1) on–off controls and (2) level controls. In its simplest form, lighting control can be accomplished manually by means of a switch located on a wall, in a luminaire or a panel box. Even though manual on–off switches for lighting control are used in commercial and industrial facilities, the current trend is toward greater use of lighting contactors. However, many circumstances require a varied level of illumination. Dimming devices are the most popular means of providing the level controls.

Lighting controls can also be grouped into two general categories: centralized controls and local controls. The main difference between them falls into the realm of function of the area considered. Centralized controls are used in buildings where it is desirable to control large areas of the building on the same schedule. An example of centralized controls is a microprocessor that turns all lights on and off on a preprogrammed schedule. Localized controls are designed to affect only specific areas. Examples of localized controls are personnel detector or a photocell controlling each office within a suite of offices.

Since a centralized system can be utilized to activate local controls, a time clock can be used to energize the building's entire lighting system on a time-of-day schedule, while a personnel detector (local control) located in a specific office overrides the centralized control to turn lighting in the specific offices on and off as demanded by occupancy. Figure 5.2 shows a

Type of Control	Total Building Control	Localized Control	Programmable	On/Off	Manual Override	Telephone Control	Time of Day	Holiday/Weekend Schedule	Dimming	Co-ordination with Daylight	Remote Control	Co-ordination with HVAC	Over-the-wire Control	Low Voltage
	Functions*													
Centralized Controls:														
Computer operated	x	x	x	x	x	x	x	x	x	x	x	x	x	x
Over-the-wire controller clock	x	x	x	x	x		x	x			x		x	x
Programmable control	x	x	x	x	x		x	x			x	x		x
Localized Controls:														
Daylight control (dimmable)[1]		x			x				x	x				x
Daylight control (non-dimmable)		x			x					x	x			x
In-fixture daylighting control		x			x				x	x	x			
Solid state dimming (no dimming ballast)		x	x		x				x	x			x	
People sensors		x		x	x									x
Ambient light co-ordinators[2]	x	x	x						x	x			x	
Timer control	x	x		x	x		x						x	

[1]Requires dimming ballast in each luminaire
[2]Compensates for light loss factors
*Many functions can be added, check with manufacturer

Figure 5.2 Lighting control matrix.

matrix of the functions required for lighting control systems, with the type of control designed to perform that function or functions. No attempt is made to differentiate the actual technology utilized by manufacturers.

5.2.3 *On–Off controls*

Wall switches

AC snap switches can be used to permit selective use of lighting and also allow for different levels of lighting in a space. A variety of devices can be used, such as three-way switches, to provide a path of lighting along a hallway. An illuminated switch is lighted when in the OFF position so it is easy to spot in a darkened room. A pilot light switch is lighted when in the ON position to serve as a visual reminder that a remote lighting load is on. A press switch allows no-hands operation so even those carrying objects can cooperate in reducing lighting energy use.

A key-activated AC switch can be operated only by a standard key, which is provided with the device, thus preventing unauthorized use of a lighting circuit.

Lighting contactors

1. Power lighting contactors — Power contactors can be divided into three general categories: power contactors rated up to 1200 A, multipole

contactors with up to 12 poles, and single-pole contactors with low voltage control. More popular sizes range from 60 to 225 A. Since small contactors can be used for control of stations and an unlimited number of stations can be used for each power lighting contactor, the illuminating engineers can be liberal with the number of control stations used. A typical installation is the Twin Towers of World Trade Center in New York City. Four 225-A power lighting contactors are used for each floor. They are, in turn, controlled by two master controllers. Being mechanically held, they can be controlled by:

   1. A manually operated three-position switch with a center-off position
   2. Auxiliary relays as a function of photoelectric cells
   3. A time switch with a single-pole, double-throw contact
   4. Control relays in an energy management system (e.g., programmable controllers)

2. Multipole lighting contactors — Multipole contactors are available with up to 12 poles with contacts usually limited to a 20-A rating. The latest multipole contactors can be provided with optical solid-state control modules that provide two-wire control, three-wire control, or stop/start control. A typical six-pole lighting contactor with a solid-state two-wire control module is shown in Figure 5.3.

Today's multiple lighting contactors are of shallow construction, permitting them to be mounted in a 4-in. stud-construction wall. Magnetically held 20-A multipole lighting contactors are also available. They are often furnished, assembled and prewired, in enclosures with electronic transceivers for connection to programmable controllers that utilize card readers, telephone interfaces, card printers, and other external components as needed.

Low-voltage relays

Low-voltage relays are usually single-pole and used for individual branch-circuit or luminaire control. These relay contacts are normally rated for a 20-A tungsten filament load at 125 V AC and mechanically latching, requiring only a momentary 24-V rectified AC switch circuit pulse to either open or close the local contacts. A step-down transformer is required to provide low-voltage power for relay actuation. With a 40 VA rating, one transformer can supply power to up to 15 relays with No. 14 AWG wiring.

The latching relay has three leads to provide the circuit path through the solenoid for latching or unlatching the line-voltage contacts. A second model has an internally energized pilot contact for a pilot light indication and so it requires four leads. A third model has an isolated internally energized pilot contact and thus has five leads. The five-lead relay with isolation of the pilot circuit is for special applications, such as wiring to a computer or a separate power source for the pilot light. In Figure 5.4 the solenoid coil is center tapped because the magnetic field strength required for unlatching is less than the magnetic field strength needed for latching.

Figure 5.3 Six-pole 20-A lighting contactor. (Courtesy of Automatic Switch Company.)

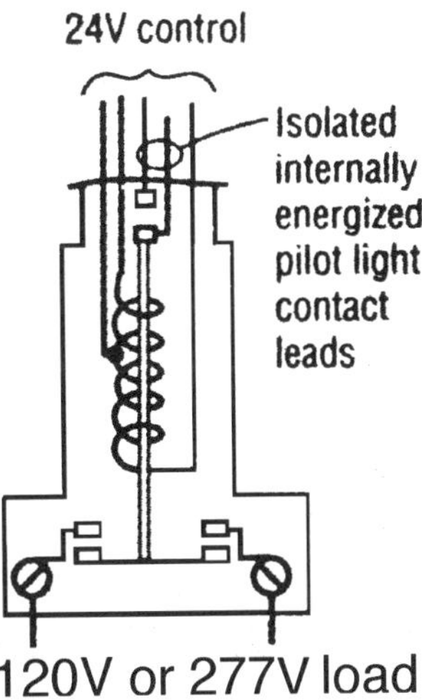

Figure 5.4 Low-voltage latching relay and wiring.

Timing controls

A broad variety of time switches are available for controlling lighting loads. They may be used for direct on–off control of lights or for control of lighting contactors. Twenty-four-hour dial timers are available with one or more sets of on–off trippers. A day-omitting device is optional for applications in which lighting is not used on weekends. Astronomic dial timers turn lights on at sunset and off at a prescribed time. Seven-day-calendar dial timers are used when the on–off program is set on one dial. Operation can be omitted on selected days. The tripper will require periodic resetting to conform to seasonal changes.

Program dial timers are used for multiple daily operations or for short-duration needs, such as turning lights on and off for cleaning crews and security guards. A total of 48 on–off operations can be programmed daily. They are available in multidial versions and with day-omitting capability. Reset timers incorporate a manual control and a timer. They are typically used by security guards on patrol; the guard can turn on selected lights when entering the area, and the timers turn off the lights after a prescribed time has elapsed.

Digital timers are used where multiple circuits, accuracy, and more sophisticated programming are required. They can be scheduled a year in advance for holidays and so on. However, they have a limited load-carrying capacity, typically 3A at 24 V AC per circuit. This requires interfacing with a lighting contactor or relay and a step-down transformer. Figure 5.5 shows the front-panel illustration of digital timers for controlling eight lighting circuits.

Phototimers combine a remotely mounted photocontrol and a 7-day dial timer. It controls three circuits, each with its own program. One circuit provides on and off operation by photocontrol; in the second, photocontrol turns the circuit on and a time switch turns it off; and in the third, on and off control are both provided by a time switch. Built-in manual bypasses maintain any circuit in on and off operation for prolonged periods without affecting the other circuits or the master program.

Sensors

Sensors for lighting controls can be divided into two categories: photoelectric sensing and presence detectors.

Photoelectric sensors — When used for control of outdoor lighting, photoelectric sensors are normally set to switch lighting on at dusk and off at dawn. Adjustments can be made to change response to higher or lower light levels. Built-in time delay helps to eliminate nuisance switching in response to sources other than natural ambient light. Ambient lighting control systems using photocell sensors can serve switching and continuous dimming functions. In response to the photocontrol unit, the switching system will turn on lights in a specific pattern to provide typically 1/3, 1/2, or 2/3 of full light output in an area. A continuously dimmable system will

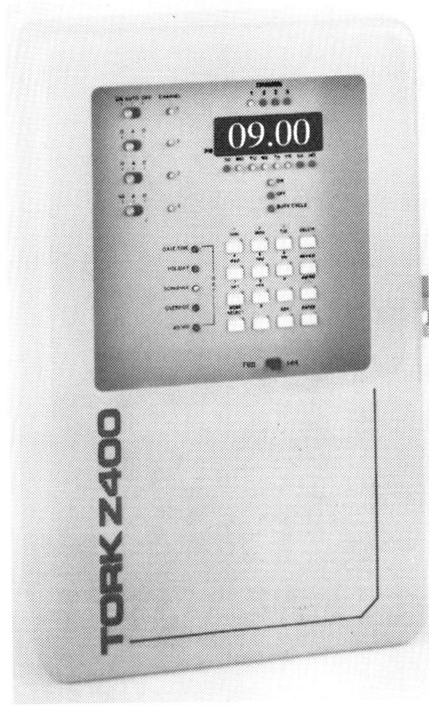

PROGRAMMING FEATURES

- Astronomic data can be conveniently copied into other channels without repeating settings
- Only need to enter latitude to set astronomic feature
- Flashing LEDs prompt the user through correct sequence of settings
- Instant entry at the touch of a key
- Positive touch keys
- Audible signals confirm each setting and notify errors
- LEDs visible from sides and distance
- Only one entry need be made for the same event on several different days
- Easy to go back and change a part of any entry
- All loads remain functioning during re-scheduling – then Z400 gives instant look back to execute the new program
- Automatically calculates day after entering date and year

Figure 5.5 Front panel of a digital timer with self-prompting LED features. (Courtesy of Tork Company.)

adjust the output of the electric lighting system in response to the amount of daylighting striking the control photosensor.

Personnel sensors — A variety of personnel sensors have recently been developed. Personnel sensors are complete control systems that sense when a space or room is occupied and automatically turn the lights on for a preset period. If no further occupancy is sensed during this time interval, the lights are then turned off automatically. There are many types of personnel sensors; they differ only in their method of sensing occupancy. Some of these methods include passive and active infrared, ultrasonic, and acoustic sensing.

Passive infrared sensors: They detect and respond to changes in radiated heat within a room caused by the presence and movement of a human body. They consist of two components, the sensor and the control unit. The sensor, which contains the passive sensing-element optics system and the electronic logic circuitry, is designed to be mounted in the ceiling panel. The control unit, which contains the low-voltage power supply for the system and the load relay used to switch the lighting load, is typically mounted above the ceiling. The sensor includes a timer circuit that keeps the lights on as long as changes in infrared energy are detected. If no changes are detected for a certain period, the lights are turned off automatically.

The most common passive infrared sensors have a coverage area of approximately 200 ft^2. If a larger area needs to be covered, multiple sensors can be connected to a single control unit, as shown in Figure 5.6, which indicates placement of three 200-ft^2 sensors for coverage of a 20 × 20-ft L-shaped office.

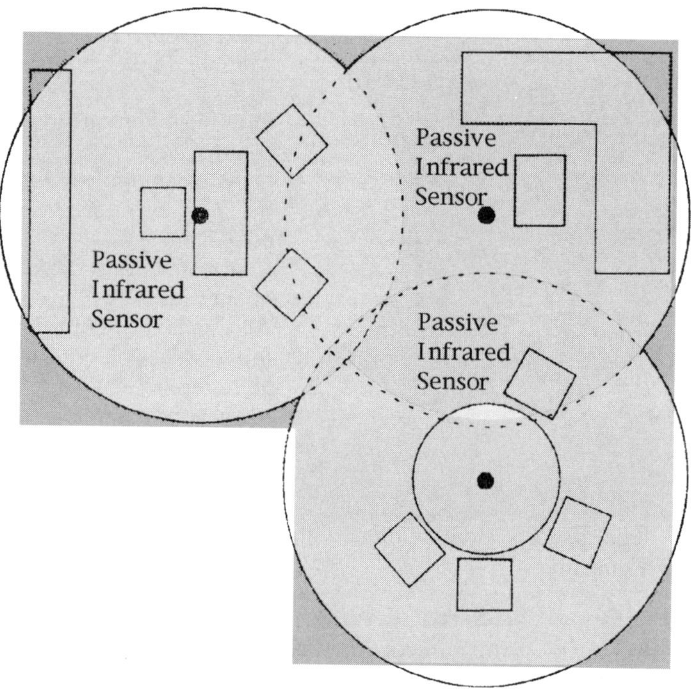

Figure 5.6 Placement of sensors for coverage of 400-ft office.

There are design considerations in the use of personnel sensors. The sensor must be placed so as to cover areas of the room where occupants are expected to be. Care must be taken to ensure that the sensors cover all potential occupant locations. The sensor should not be placed too close to the entry; otherwise, people walking in an outside corridor past an open door will activate the lights. Passive infrared sensors should also be placed where they will not sense any nonhuman heat sources, such as an HVAC register or baseboard heater.

Ultrasonic sensors: An ultrasonic sensor creates an ultrasonic field in the room being monitored. When a person enters the room, the field is disturbed; the sensor detects the disturbance and activates the room lights. A typical ultrasonic sensor unit containing all the sensing and local control equipment is mounted on the ceiling and can provide up to 900 ft^2 coverage. A single unit often contains multiple sensors.

Each type of ultrasonic sensor has a coverage pattern that depends on the number of sensors mounted in the unit. The pattern may include two different coverage areas for the same unit: an area within which large body motions, such as walking, are detected, and a smaller coverage area, within which small body motions, such as the movement of an arm, are detected. Careful placement of the ultrasonic sensors ensures proper operation of the system. Sensors must be placed such that they do not detect motion outside the room being controlled. Another consideration in the application of ultrasonic sensors is the acoustics of the room. Room acoustics can affect the coverage pattern and therefore the appropriate number and placement of sensors.

Ultrasonic sensors are available in many different styles, each providing specific coverage patterns. There are also ultrasonic sensors designed for wall mounting as a direct replacement for a doorway light switch. Personnel sensors are ideal for controlling lights in any space with random and intermittent occupancy patterns. Some spaces that meet this criterion include enclosed offices, rest rooms, and storage rooms.

Acoustic sensors: They respond to sounds created as a result of a person's movement in an area. The unit is useful where direct line of sight to a sensor cannot be achieved, such as an irregularly shaped room or corridor, a stairway, or a storage room.

Active infrared sensors: They emit invisible infrared beams in a specific pattern and a receiver responds to changes in the light beam patterns caused by a person's movement in the field of view.

Programmable control systems

Control systems are now available that employ microprocessor logic to replace hard wiring with soft wiring. Coded comments can be multiplexed to control points over a pair of low-voltage wires. Control points have receiver/switches, the latter complement generally a low-voltage relay or lighting contactor. Logic functions can be programmed into a control device to turn lights on and off over a 24-h or 1-week period. Overrides are available, and some systems can be accessed with touchtone telephones.

Such systems, which control lighting in both time and space, save considerable energy compared with past control practices. The heart of such systems is the programmable controller, which holds in memory a series of on–off instructions to the lighting circuits throughout a building. They can provide minute-by-minute control of an entire lighting scheme according to a user-determined schedule, with pulse initiation of the control signal generated from its internal clock.

Control mediums

A number of methods are used to provide area-wide control in a building:

Centralized programmable control — Multiplexed signals from the programmable controller can be sent through a building via twisted low-capacitance wire. Signals can also be sent via existing power wiring to the receiver control modules. The codes from the controller are transmitted to the transceiver, where they may be combined with other signals (e.g., photocontrol relay output). In turn, signals are provided to low-voltage relays or lighting contactors, which require only a momentary electrical pulse to operate its latching or mechanical mechanism to the on or off position.

Control of the lighting pattern is not limited to the programmable controller keyboard. In addition, a manually operated, momentary contact switch connected to the transceiver can provide signals to operate not only those relays, or contactors connected to it, but also relays or contactors located anywhere throughout the building via the data line back to the programmable controller.

Power-line carrier system — The power-line carrier system contains electronic transmitters and receivers that use the building power wiring as a communication pathway. The transmitter accepts a control signal input, converts the control signal into digital form, and injects it onto the power wiring system. The low-voltage digital signal is transmitted at a frequency anywhere from 25 to 250 kHz, depending on the specific transmitter design. It is important to note that host signals cannot pass through transformers; bypass devices are usually required at each transformer.

Radio-controlled ballast system — One interesting use of power-line carrier technology is a high-intensity-discharge (HID) dimming ballast that has a receiver built into it. Control signals carry dimming information to all ballasts equipped with integral receivers. This system can be connected to a photosensor to provide for either daylighting compensation or lumen maintenance control. A power-line carrier system can be economically used in large facilities with many control points, such as factories and warehouses where control cabling would be lengthy and expensive. The technology is also suitable for existing buildings, where the cost of installing new lighting control wiring can be prohibitive.

5.2.4 Level controls

Dimmers

Many circumstances require a varied level of illumination. Dimmers are the most used means of providing lighting level control. The original dimmers were of the resistance types. In the past decade, solid-state dimmers have taken over 90% of the market. By means of an electronic switch, the electronic dimmers turn off the current to the load for a portion of the cycle, thus delivering less power to the load. Electronic dimmers are now available for incandescent, fluorescent, and HID lighting.

Types of dimmers include:

1. Conventional dimmers — The modern SCR (silicon-controlled rectifier) dimmer operates on the principle of switching on the current a proportional distance through each half-cycle. An SCR is nothing more than a very fast switch. Dimmers operate simply by delaying the turning on of these switches by an amount of time inversely proportional to the incoming control voltage. In the simplest of examples, a dimmer set at 50% will delay halfway through each half-cycle and then turn on.
2. Digital dimmers — It uses the same very fast switch (SCR), but instead of using a voltage to charge a capacitor and turn on the switch, the digital dimmer counts a number of steps through each half-cycle, then turns on the switch. Another term that is often used when referring to new technology dimmer is multiplexing. Multiplexing refers to the use of a single cable to carry data to a group of dimmers. Multiplexing is a concept that can be used on both analog and digital dimmers.
3. Intelligent dimmers — They communicate with the controller in the same way as does the digital dimmer except that they have the intelligence to recognize data other than the setting of the controller. The scope of what these data could include is as broad as the scope of what the intelligent controller can produce.

Technology of patching

As dimmer per lighting system becomes more and more clearly the wave of the future, moving the job of patching out of the controller and into the dimmer makes more and more sense. If in this data stream the intelligent dimmer is being informed that it is patched to channel 5 and that whenever it sees data for channel 5 it should use them, the number of dimmers that can be patched to any control channel will be limitless. One problem when lights are patched together in a single control channel is that the light produced when that channel is active may not be uniform across an area. If a proportional level is a part of the data that the intelligent dimmer stores and uses, all the dimmers patched to one channel can be balanced to make the lighting uniform. It is possible for an intelligent dimmer to communicate information gathered back to the controller so that the controller can make adjustments and/or communicate this information to the operator. For years, conventional analog dimmers have used feedback information to adjust the operation of dimmers — line regulation, load regulation, and current limiting being most common.

The intelligent dimmer will be able to do as many of these functions as engineers deem desirable. The intelligent dimmer will run a program like any other computer, and this program can be changed. The scope and capabilities offered by it are going to be important parts of control technology in the future.

Computer/microprocessor

Most sophisticated lighting control now consists of a microcomputer, an oscillator as its controller, and a receiver switch as decoder. The microcomputer is programmed for various lighting patterns and addresses of the luminaires. The address and condition codes generated are modulated by the oscillator and superimposed on the building electrical system line frequency. At the receiver/switch, a decoder takes the message off the line, and if the address code corresponds to the one given that switch, the condition codes are executed and turn the luminaires fully on, halfway on, or off. One such system contains all programming, logic circuitry, and hardware. Discretionary control and override functions can also be incorporated. Printer can be connected to the microcomputer to provide hard-copy records of all control activities. This feature gives information to the system operator about the frequency and time of any local override activities. Video display terminals and recent software have made the system much easier to use, which promotes prompt revision of control schedules. Transceivers are available in 16- and 32-output versions. The control relays are capable of switching a 20-A inductive load. Figure 5.7 shows a typical scheme for such a control system as just described.

Manual incandescent dimming devices

1. Wallbox-type dimmers — They are rated up to 2000 W of lighting load to be manually controlled from a single electrical device box. Some models allow dimming from two or more locations, which previously required a remotely controlled dimmer module. Wallbox dimmers are also available with integral electronic switching to allow remote on/off control from multiple locations, using controls that aesthetically match the dimmers.
2. Microprocessor-based wallbox dimmers — They combine dimming control, preset memory of a desired lighting level, and time clock operation. A typical unit containing four dimmers, each with four scenes of possible settings, fits a standard multigang device box.
3. Console-sized, microprocessor-based systems — They can provide memory for hundreds of scenes and lighting zones, and the automatically produced dimmer settings can change at very slow fade times.

5.2.5 Bases for selection of lighting controls

Guidelines for minimum number of lighting controls

Table 5.1 represents an attempt to establish some guidelines for a minimum number of lighting controls to be installed with a reference to the area size and unit power density of a lighting system. The controls used are either on–off devices or dimming controls.

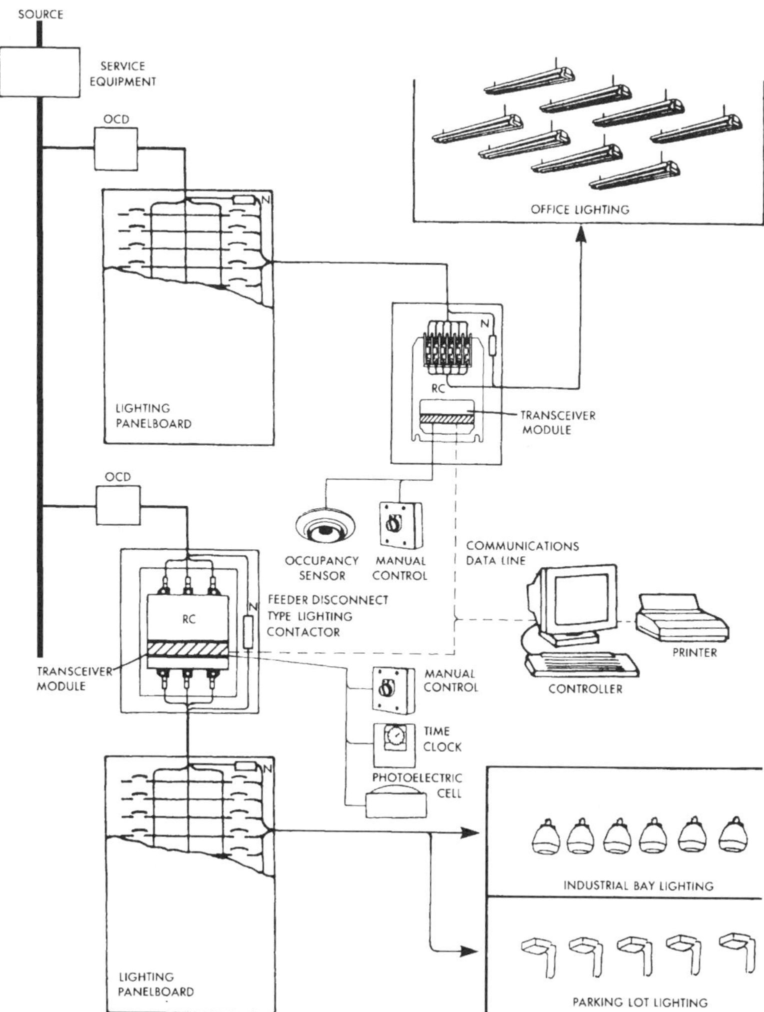

Figure 5.7 Programmable lighting control scheme. (Courtesy of Automatic Switch Company.)

Medium-sized offices

An assessment of the cost effectiveness of photoelectric control equipment was made for three medium-sized offices:

1. A fully automatic mixed control system would be unlikely to be cost effective for single offices.
2. A partially automated on–off system would be cost effective in new buildings at present energy costs.
3. A partially automated mixed system for new buildings would probably be cost effective only if designed to control luminaires in several offices.

Table 5.1 Recommended Minimum Number of Lighting Controls[a]

	UPD (W/ft²)		
A (ft²)	1.5	1.5–3	3
125	1	2	2
125–250	1	2	2+
251–500	1	2+	4
501–1000	2	3+	5+
1001–2000	3	4+	6
2000	3+	5+	6+

a. When multiple controls are used, it is generally installed to permit reducing the general lighting in the space by at least one half in either a uniform pattern or by zones, as most appropriate.

The results of this study are summarized in Table 5.2. From a close examination of Table 5.2 it would seem sensible to include a dimming line to each luminaire during its installation in new buildings. This would involve only a small extra capital cost, but would keep open the options of installing a dimming system later without the necessity for complete rewiring. In another study, when due considerations were given to daylight factor, it appeared that it is generally best to control only those luminaires nearest the windows.

Table 5.2 Cost Effectiveness of Photoelectric Control Systems[a]

	New buildings		Existing buildings	
Type of control	Single office	Multi-office	Single office	Multi-office
Fully automated, 1 on–off + 1 dimming	2	1	2	2
Partially automated, 1 on–off	1	1	3	3
1 on–off + 1 dimming	2	1	2	3

a. 1 — Cost effective within 15 years; 2 — not cost effective within 15 years; 3 — depending on energy costs.

Factors affecting selection of lighting controls

In general, there do not appear to be any general rules or guidelines that conveniently lead one to select specific controls. The following factors will have a bearing on the selection of lighting controls:

1. Size of facility — A large facility may justify a building management system or programmable controllers that provide centralized lighting control. On the other hand, a small facility may obtain optimum

savings by selecting a simple time switch control. From experience, large computers often cost around $500 per hour to operate, whereas a minicomputer costs no more than $10 per hour to operate.

2. Size of individual lighting area — If the current drawn by an individual lighting area exceeds 20 A a power contactor may be the choice. Small areas such as individual offices would be candidates for low-voltage relays. In either case, these devices would be controlled by timers, photoelectric sensors, and the like. In recent years, the trend has been to control smaller individual lighting areas.
3. Availability of daylighting — As discussed previously, energy savings from daylighting depends on many factors: climate conditions, building form and design, and the activities in the building. Only a portion of a building can be daylit; however, in most cases, 30% of the floor space is sufficiently close to the perimeter to be daylit. There is little documented research in this field. As a general guide, the average energy savings from daylighting for an entire building will be in the neighborhood of 15%.
4. Type of usage in the facility — If the facility provides commercial rental of office space, consideration is often given to flexible controls, such as low-voltage relays. In the institutional facilities where lighting requirements are more fixed, other types of controls should be considered.
5. New installations or modification of existing facility — From Table 5.2 it becomes evident that a more elaborate photoelectric control system or even a sophisticated lighting management system can be easily justified for a new building. However, for an existing building, an extensive lighting control system may not be cost effective. In this case a control system that utilizes existing power wiring for signal transmission would be preferred.

Comparison of lighting control systems

Table 5.3 shows a simple comparison of different systems according to how well each meets the needs of both the occupant and the building management. The relative performance rating of each system is often a judgment call. Individual manufacturers for a single system may vary sharply in system performance and cost. The objective here is to process, not pass final judgment or provide pricing guidelines. Some of the reasoning that went into the relative rating for each function shown in Table 5.3 is given below:

1. Occupancy sensitivity — From an occupant's perspective, individual wall switches work just fine. Contactors in most building automation systems can be a real disadvantage, since they do not normally allow the occupant to override for after-hour usage. Programmable lighting control caters to the occupant. When he or she enters the building after

Table 5.3 Comparison of Various Lighting Controls

	Occupant Sensitivity	Light level selection	Energy savings potential	Management data	Intergration capability	Space adaptability	Cost
Standard wall switches (indiviudal offices)	Good	Fair	Fair	No	No	Poor	Medium
Contractor control via building automation system	Poor	Poor	Poor	Yes	Yes	Good	Low
Programmable lighting control relay based	Excellent	Fair	Good	Yes	Yes	Good	Medium
Programmable lighting control dimming based	Excellent	Excellent	Excellent	Yes	Yes	Good	High
Occupancy sensor	Fair	Poor	Good	No	No	Poor	High

hours, the person's particular working area can be lit in anticipation of his or her arrival with a single phone dialing. Similarly, when a person is staying late, his or her office and related work space can both be kept on with a phone dialing or switch override.

2. Occupant-level selection — This is a function affected by both control system capacity and floor layout. Individual office layouts with manual switching of split-wired fixtures or several lighting sources within the space give most occupants the degree of control they need. The dimmable solid-state ballast approach provides an even better method for allowing the occupant to adjust the overhead lighting.
3. Energy-saving potential — The only surprise here is in the "good" rating for switches in individual offices and the "poor" rating for the same devices used to control a zone. In practice, what happens is that when more than one person is in a zone, the first in turns it on and the last out never looks back.
4. Management data — These reflect system monitoring analysis, and reporting capabilities. These require a communications capability not normally inherent in switches or occupancy sensors.
5. Space adaptability — The important point here is that devices physically linked to the occupant's walls or ceiling pose problems when it is time to rearrange a space.
6. Costs — The solid-state ballast cost represents a combination of functions not presently available on the market. Perhaps the biggest surprise is that individual office switches are not cheap. They typically cost $0.50 per square foot. Occupancy sensors may reduce the installation labor, but the added hardware content still means a relatively high total cost. Both switches and sensors incur added cost for office rearrangements.

In conclusion, regardless of the type and/or size of the facility which illuminating engineers may deal with, there will always be a suitable type of control for them to choose from. The engineers must become knowledgeable in lighting controls and exercise their sound judgment in the selection of control schemes to achieve a high-quality lighting system with optimum energy savings.

References

Alling, W.R., The integration of microcomputer and controllable output ballast — a new dimension in lighting control, *IEEE Trans. Ind. Appl.*, IA-20(5), September/October, 1984.

Chen, K., and Castenschiold, R., Selecting lighting controls for optimum energy savings, *IEEE-IAS Annu. Conf. Proc.*, October, 1985.

Clanton, N., Daylighting increases productivity while cutting costs, *Energy Users News*, November, 1996.

Crisp, V.H.C., Preliminary study of automatic daylighting control of artificial lighting, *Light. Res. Technol.*, 9 (1), 1977.

EC&M, Lighting Controls Help Reduce Energy Consumption, pp. 61-64, November, 1990.

IES RP-5 Recommended Practice of Daylighting, Illuminating Engineering Society of North America, New York, 1979.

Nelson, K., Daylighting benefits extended beyond just energy savings, *Energy Users News*, February, 1997.

NLPIP Specifier Reports on Occupancy Sensors, Rensselaer Lighting Research Center, Troy, NY, October, 1992.

Pearlman, G. W., The emergence and future of intelligent dimmers, *Light. Des. Appl.*, pp. 22–23, June, 1982.

chapter six

Energy-efficient illuminating system components—light sources, ballasts, and luminaires

6.1 A gaze into the future of lighting system components

The future of lighting holds the development of more compact lamps and continued research into reduced wattage incandescents, optimization of fluorescent lamp colors, large-diameter acrylic fiber "light pipes", and the induction type of discharge light. The continuing development of more compact sources in high-intensity discharge (HID), halogen-sourced incandescents, and fluorescents is leading to greatly improved optical systems.

In the HID field, it is now practical to envision pencil-thin arc tube, with resulting reduced-sized envelopments. The trend to T8 fluorescents with continued development of more efficient phosphors will also lead to much improved optical systems. The use of reflective film on the interior of incandescent lamps to reflect the heat portion of the filament output back on the filaments, causing it to operate at a higher temperature per watt of input, has resulted in much improved lamp efficacies. Continued research in phosphor technology will bring more optimization of fluorescent lamp colors. Research into large-diameter acrylic fiber "light pipes" will continue. A 60Q light source requiring 72 W input has been shown delivering 3000 lm into the pipe, which has a diameter of 1 in. The "pipe," which has losses of approximately 1% per linear foot, can be bent around the corners, with many potential future applications. Research also continues in the induction type of discharge light where there are no internal electrodes and the fill gas is excited through the field of high-frequency energy of a surrounding antenna. In fluorescent lamp circuits, electronic ballast has reduced losses almost to their practical limits. Improvements in power factor to the 97 to 99% range, and in total harmonic distortion to less than 10% are around the corner.

In the electronic ballast, it is expected to see more plug-in-type ballast connections, which have been popular in Europe for a number of years. The use of such connections simplifies manufacturing of fixtures, makes field maintenance more economical, improves reliability, and leads to a higher degree of safety.

6.2 Light sources

Light sources are generally classified into incandescent, fluorescent, and HID types. They differ considerably in physical dimensions, electrical characteristics, spectral power distribution, and operating performance. Some are better suited than others to certain applications; however, sometimes two or more sources may qualify to fulfill a specific lighting requirement.

6.2.1 Incandescent filament lamps

Initial efficacy of typical incandescent lamps (25 to 1000 W) ranges from 10 to 23 lm/W. They are commonly designed for approximately 1000 h of life. It is generally known that incandescent lamps conform to the supply voltage. A change of only a few volts can seriously affect both life and light output. Several special types of incandescent lamps are discussed in the following:

1. Reflectorized (R, PAR, and ER) lamps — These lamps have self-contained reflectors and are manufactured in a number of sizes, from 30 to 1500 W, and in various light distributions. In general, these lamps have a better-maintained illuminance. ER lamps (50, 75, and 120 W) control their beams such that they focus about 2 in. in front of their faces. Especially useful in "baffled downlight" luminaires, they permit high light utilization with attendant savings in energy. Figure 6.1 shows an ER30 lamp and its beam focus.

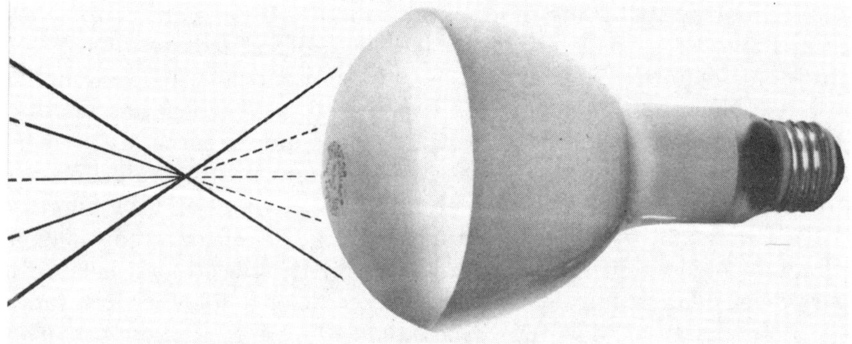

Figure 6.1 An ER30 lamp and its beam focus.

2. Spotlight, floodlight, and projector lamps — Characteristic features of all lamps designed for spotlight, floodlight, and projection applications are compact filaments accurately positioned with respect to the base, for the purpose of light control; relatively short life, for high efficacy and luminance; comparatively small bulbs and restricted burning position. Some projection lamps for use in certain types of projectors have an opaque coating on the top of the bulb to prevent the emission of stray light. Low voltage PAR lamps designed at 5.5 and 12 V produce a concentrated beam pattern with good optical control. They can be used in the fixture which may be recessed, surface, or track mounted.
3. Tungsten-halogen lamps — These lamps employ halogen to preclude blackening of the tubular envelope. They have extremely good lumen maintenance over a life of 2000 h or more. The shape of the lamps enables the luminaire to provide excellent beam control. These lamps are both single and double ended. They may be sealed into outer bulbs such as PAR types for good optical control. Because of their characteristics, tungsten-halogen lamps find extensive application in floodlighting, aviation, photographic, photocopy, special effects, and special application lighting. They have also found broad use in automobile headlighting.
4. Infrared lamps — Infrared lamps used in the home and for therapeutic purposes are commonly of the convenient self-contained 250-W R40 bulb with internal reflector and red bulb face. Those used in industrial processes are of three types: reflector, clear G-30 bulb, and T3 quartz bulb. Gold-plated or specular aluminum reflectors are most effective for use with unreflectorized infrared lamps.
5. High- and low-voltage lamps — These are available in 100 to 1500 W for 230- and 250-V circuits. They have less rugged filament, require more supports, and are less efficient than are 120-V lamps of equal wattage. Lamps for operation on 30- and 60-V circuits are also available for use in train lighting and in country home service.
6. Special lamps for industrial applications — (1) Rough service: Rough service lamps (25 to 500 W) are made with extra filament supports to withstand mechanical shock and are used principally with extension cords; (2) Silicone-rubber-coated lamps: These lamps have special rubber-like coatings that serve to reduce breakage from both thermal and mechanical shock; or should breakage occur, the glass fragments nearly always remain intact. Available in sizes from 25 to 200 W, they are especially suited to food-packaging industries and to others where manufacturing functions may subject lamps to mechanical damage; (3) Extended service lamps: These lamps operate for approximately two to three times the normal rated life. They are useful where cost of lamp replacement is high and cost of power is low; (4) Thermal shock-resistant lamps: These are available in various wattages and bulb shapes and are recommended for applications where moisture may fracture the hot tub.

7. Energy-efficient incandescent lamps — Energy-saving potential exists where reflector lamps can be applied. A new line of indoor reflector lamps (ER) allows reduction of 50% or more in energy consumption in many installations.

Incandescent lamps with special radiation-covering envelopes are being developed. Compact high-emissivity filaments are required to absorb infrared radiation reflected from specially coated envelopes. The energy-saving potential for the new lamps could be as much as 60% compared with equivalent conventional incandescent lamps. More recent developments are the application of a tungsten-halogen cycle and the use of selective reflecting thin film. As already discussed in number 3 above how the tungsten-halogen cycle works, the selective reflecting thin film comprises many layers of a particular thickness of the order of the wavelength of light. The film is transparent to visible light and highly reflective to the near- and far-infrared radiation (IR) greater than 7000 A. The bulb wall is shaped such that the IR is reflected back onto the filament, further heating the filament. The technique increases the lamp efficacy by a factor approaching two. As a result of this development, the halogen-infrared lamps are produced. They operate at wattage lower than standard halogen lamps. One lamp manufacturer claims as high as 20 LPW efficacy for its halogen-infrared PAR lamps.

6.2.2 Fluorescent lamps

The fluorescent lamp is an electric discharge source in which light is produced by the fluorescence of phosphors activated by ultraviolet energy from a low-pressure mercury arc. The lamp requires a ballast to limit the current and, in many instances, to transform the supply voltage. Lamp performance is influenced by the character of the ballast and luminaire, line voltage, ambient temperature, burning hours per start, and air movement. Fluorescent lamps are available in many variations of "white" and in a number of colors. Standard "cool white" has been most popular for industrial lighting. The efficacy of cool white lamps varies between 30 and 100 lm/W (exclusive of 20% power loss in the ballast). Although most fluorescent lamps have tubular envelopes, there are special types, such as circular, U-shaped, reflectorized, and jacketed.

Energy-efficient fluorescent lamps: Since the 1970s there has been a line of reduced-wattage replacements for standard fluorescent lamps. These lamps are now available in all popular sizes and colors for most applications. Limitations of energy-saving reduced-wattage lamps are

1. Should be used where ambient temperature does not drop below 60°F
2. Should be used only on high-power-factor ballasts
3. Not to be used where drafts of cold-air ducts would be directed

The following are several of the most energy-efficient fluorescent lamps which are commonly used today.

Compact fluorescent lamps

A recently developed arc-discharge lamp, called compact fluorescent lamp (CFL), can replace an incandescent light source in particular applications. Three configurations are possible for the installation of CFLs: dedicated, self-ballasted, and modular. Dedicated CFL systems are similar to full-size fluorescent lighting systems in which a ballast is hard wired to lampholders within a luminaire. Self-ballasted and modular CFL products have screw base sockets; they typically replace incandescent lamps. A self-ballasted CFL contains a lamp and ballast as an inseparable unit. A modular CFL product consists of a screwbase ballast with a replaceable lamp. The ballast and lamp connect together using a socket-and-base design that ensures compatibility of lamps and ballasts. While most of the modular types are operated in the preheat mode, the electronic ballasted lamps are operated in the rapid start mode and in principle could be dimmed.

CFLs are developed as an economical substitution for lower-wattage incandescent lamps in applications requiring long burning hours, such as corridors, stairwells, lobbies, and reception areas. The economics are two-fold: the CFL has a rated life of about 10,000 h and even when ballast losses are included, the CFL offers four times the efficacy of an incandescent lamp. In addition, being the first lamp type to use the triphosphor coating, the CFL has a CRI of 80 or better and a 2790 K color temperature, making it compatible in appearance with the chromaticity of an incandescent lamp.

CFLs have either a T-4 (10 mm) or a T-5 (15 mm) glass envelope that is bent in a U shape and mounted on a special base. Ratings are 5, 7, 9, 13, 18, and 26 W. The single twin-tube CFL is too long for some applications, so the family of double twin-tube or "quad" tube lamps was introduced. The recent new products include triple lamps up to 42 W and a 40-W lamp bent into a helical shape that looks like a Dairy Queen cone. In many cases, the CFL should be matched to a specific ballast type from a particular manufacturer. If a CFL is compatible with a particular ballast, it may not properly start and operate. Figure 6.2 shows a modular CFL which is a double-folded, bent-tube assembly that can be retrofitted to a standard medium-base socket.

T8 lamps

Although low-price 34-W T12 cool white lamps are available to replace the 40-W T12 cool white eliminated by the Energy Policy Act, these lamps have higher lumen depreciation, are more sensitive to low temperatures, are not appropriate for dimming, and have a potential lamp-ballast compatibility problem. Instead, the 32-W T8 lamp utilizing rare earth phosphors has both high efficacy and good color rendering. It has become the standard for new construction and is increasingly popular as a retrofit replacement for 40-W T12 lamps.

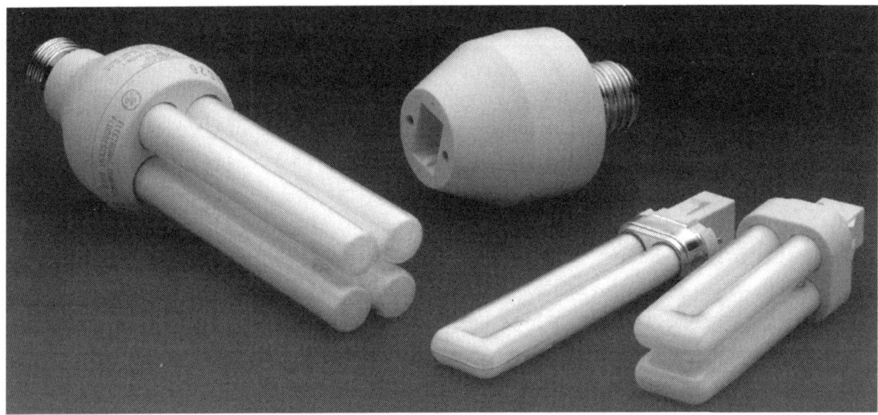

Figure 6.2 A modular compact fluorescent lamp assembly.

T5 lamps

T5FT fluorescent lamps are distinguished by their shape and high light output compared with CFLs and linear fluorescent lamps. Because of their small diameter and compact shape, controlling light distribution is easier with T5FT lamps, and luminaire efficiency can be improved. T5 lamps are slightly shorter than U.S. standard linear fluorescent lamps. Presently only a few ballast and luminaires are available for T5. But T5 lamps may become popular because of their high efficacies and small diameter.

Electrodeless lamps

These lamps use a high-frequency electromagnetic field to directly excite the gas fill in the lamp so that electrodes are not needed. Lamp life can be much longer. Currently there are two different types: inductive discharge lamps and microwave-discharge lamps, such as sulfur lamps. The sulfur lamp is a higher power 1375-W system with very high light output of 130,000 lm and with a rated life of 15,000 h for the magnetron and 45,000 h for the lamp. There will be no spectral shift over the life of the lamp. It has a high color temperature of 6000 K. One of the concerns for the electrodeless lamp is electromagnetic interference (EMI). One lamp manufacturer claims that its lamp will meet the FCCs EMI requirements for commercial applications.

6.2.3 High-intensity discharge lamps

High-intensity discharge (HID) lamps are electric discharge sources. The basic difference from fluorescent lamps is that HID lamps operate at a much higher arc pressure. Spectral characteristics differ from those of fluorescent lamps because the higher pressure arc emits a large portion of its visible light. HID lamps produce full light output only at full operating pressure usually several minutes after starting. Most HID lamps contain both an inner and an outer bulb. The inner bulb is made of quartz or polycrystalline aluminum; the outer bulb is generally of thermal shock-resistant glass. HID lamps require current-limiting devices, which consume 10 to 20% additional power. HID lamps include mercury, metal halide, high-pressure sodium, and low-pressure sodium lamps.

Mercury lamps

They are low in efficacy compared to other HID sources, and are obsolescent for most industrial applications. They are available with either "clear" or phosphor-coated bulbs of 40 to 1000 W, and in various sizes and shapes. Typical efficacy ranges from 30 to 63 lm/W not including ballast loss. "Clear" mercury lamps produce light rich in yellow and green tones while lacking in red. Phosphor-coated lamps provide improved color. Special types include semireflector, reflectorized, and self-ballasted lamps.

Metal halide (MH) lamps

These are similar in construction to mercury lamps. The difference is in the arc tube, which contains various metal halides in addition to mercury. They are available in either clear or phosphor-coated bulbs from 32 to 1500 W. Present efficacies range from 70 to 125 lm/W, not including ballast power loss. Color improvement can be achieved by the additives.

Recently a new technology — using the arc tube material of high-pressure sodium lamps but with metal halide chemistry — results in warmer, more incandescent-like color properties than ever before. From this new technology come several families of compact metal halide lamps, especially PAR20 and PAR30 display lamps. Compact, light-weight electronic ballasts enable 35- and 50-W lamps to be mounted to track lighting systems in attractive luminaires. A 35-W PAR20 appears to perform as well as the most efficient halogen lamp operating at 100 W or more. If prices on ballasts and lamps become competitive, this could be a breakthrough for many applications.

High-pressure sodium lamps

Light is produced by electricity passing through sodium vapor. They are presently available in sizes of 35 to 1000 W. Typical initial efficacies are about twice that of mercury lamps: from 80 to 140 lm/W, not including ballast power loss. Normally with clear outer envelopes, they may also be obtained with coatings that improve diffusion. The color of light produced is golden white. Figure 6.3 shows several 250-W high-pressure sodium (HPS) lamps.

For interior applications, lamp manufacturers have made major advances in improving the color of HPS lamps. Although the improved-color lamps have lower efficacies than standard HPS, their efficacies are higher than incandescents. They can be used in downlights, recessed and semirecessed adjustable accent lights, and track luminaires. Color stability over lamp life is a major advantage compared with low-wattage metal halide lamps.

Figure 6.3 250-W high-pressure sodium lamps.

Low-pressure sodium lamps
These are presently available in 35 to 180 W. Typical initial efficacies are high: 137 to 183 lm/W, exclusive of ballast power loss. Applications are limited by virtue of their monochromatic yellow color. Figure 6.4 shows a typical low-pressure sodium lamp.

6.3 Ballasts

6.3.1 Fluorescent ballasts

Most fluorescent lamps operate on one of three types of ballast circuit: preheat, instant start, or rapid start (Figures 6.5 to 6.7). A few can be operated with either preheat or rapid-start ballasts. Preheat lamps up to 20 W can be operated on special rapid-start (trigger-start) ballasts. Preheat lamps operated on preheat ballasts require auxiliary "starters" to allow current to flow through the electrodes for a few moments before the arc is established across the length of the lamps. Instant start and slimline lamps require no starters. The ballasts provide enough voltage to light the lamp instantly. Rapid-start lamp operation is also starterless. Rapid-start lamps are most popular for new fluorescent lighting installations. Fluorescent lamp ballasts are available for most secondary distribution voltages.

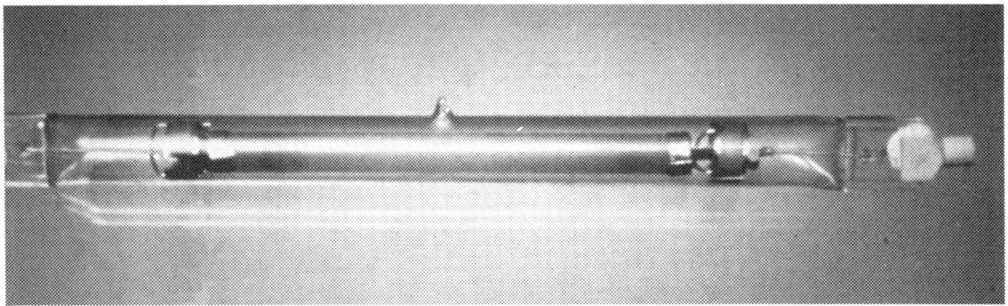

Figure 6.4 Typical low-pressure sodium lamp.

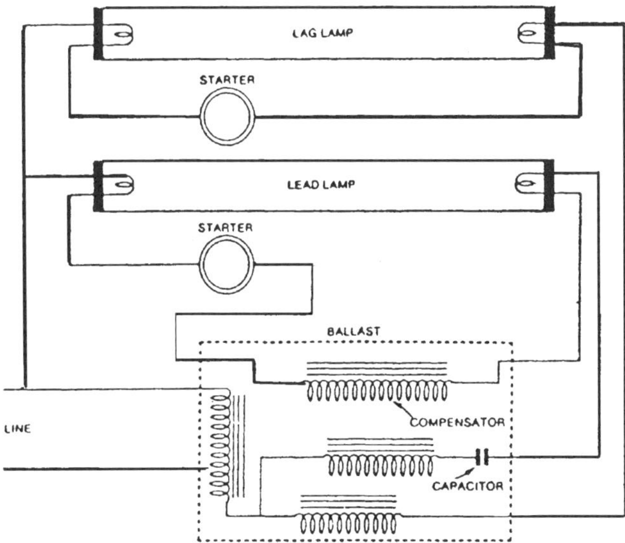

Figure 6.5 Preheat fluorescent ballast circuit.

Energy-efficient ballasts

In general, the greater the ballast size (power rating) the higher is the ballast efficiency. A two-lamp ballast is more efficient than a one-lamp ballast. The ballast efficiencies can generally be calculated from the manufacturers' catalog data. The internal ballast losses are determined by coil construction, the nature of the magnetic materials, and resistance of the conducting coil wire. With increased energy costs, ballast manufacturers have introduced energy-efficient ballasts that minimize the ballast losses. The following details some of the new ballasts:

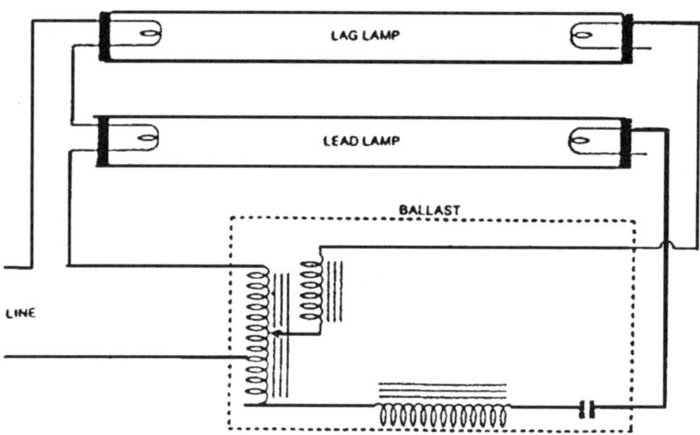

Figure 6.6 Instant-start fluorescent ballast circuit.

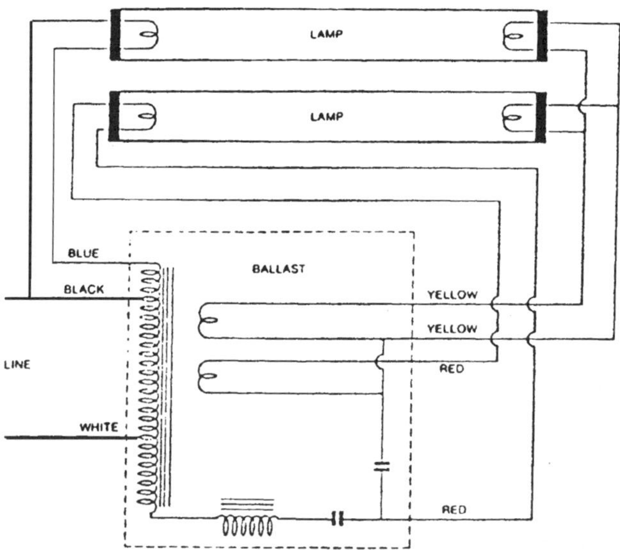

Figure 6.7 Rapid-start fluorescent ballast circuit.

1. Low-energy ballasts — Fluorescent lamps may be operated at less than rated power and output, provided starting voltage and operating voltage requirements are met. In the case of rapid-start lamps, cathodes shall be heated regardless of the current through the lamp to provide rated lamp life. Low-energy ballasts are low-current designs which can provide energy reductions compared to standard units. They are useful where certain luminaire spacing to mounting height criteria shall be followed, and the desired illumination level is less than that obtained by full output operation of the lamp. Before operating fluorescent lamps

at less than rated output, it is wise to check the ambient temperatures of the air surrounding the lamps. Higher minimum ambient temperatures are required when low-energy lamps or standard lamps are used with low-energy ballasts.

2. High/low ballasts — This type of ballast, which is generally available only for rapid-start circuit operation, contains extra leads which can be connected or switched to provide multilevel operation of the lamp. Two- and three-level rapid-start ballasts are available, and fixture output may be set according to the lighting requirements of the area. Operation of fluorescent lamps at less than rated output may raise minimum operating temperature requirements.
3. Low-loss ballasts — Ballasts may be designed to reduce internal losses by improving mechanical and electrical characteristics. More efficient magnetic circuits, closer spacing of coils, and improved insulation systems can result in loss reductions of nearly 50%, compared to conventional units. The ballast manufacturers' ratings should be consulted to determine which ballast will have the least losses for the power system and lamp combination involved.

Electronic ballasts

The efficiency of fluorescent lamp systems can be improved further by utilizing electronic ballasts. Improvements occur two ways:

1. By way of lower internal losses within the ballasts
2. By making use of the fact that fluorescent lamp efficacy increases as a function of the frequency of the applied power

Electronic ballasts usually operate a lamp at frequencies above 20 kHz. In general, operating a fluorescent lamp with an electronic ballast will realize a 20 to 25% increase in system efficiency compared to a standard core-coil ballast. The electronic ballast offers improved performance which includes: (1) reduced flicker, (2) improved voltage and thermal regulation, and (3) the ability to dim over a wide range without affecting lamp life. Table 6.1 shows the performance of the standard and energy-efficient core-coil ballasts and also the electronic high-frequency ballasts.

The data given are for two lamp systems. The 40-W F40T12 cool white lamp is still in general use; however, the 32-W F32T8 lamp is gaining in popularity based on its high system efficiency. The table shows that the electronic ballast can meet all necessary lamp parameters to maintain lamp life. The high harmonic data for the T8 electronic ballast were designed before harmonics became an issue. Today the technology is available to reduce the harmonic contents within a specified limit; the constraint would be the cost or economic considerations. Electronic ballasts are available for all of the commonly used fluorescent lamps, including 34-W F40T12, 40-W F40T10, and 32-W F32T8 (rapid and instant start), all of the F96-type lamps. There are one, two, three, and four lamp ballasts. The new generation of

Table 6.1 Performance Data of the Standard and Energy-Efficient Ballasts Vs. Electronic Ballasts on Fluorescent Lighting Systems

Ballast lamp	EE Mag 40-W F40 T-12 CW	EE Mag 40-W F40 T-12 CW	Electronic Ballasts 40-W F40 T-12 CW	Electronic Ballasts 32-W F32 T-8 41 K	Electronic Ballasts 32-W F32 T-8 41 K
Power (W)	85	81	74	69	65
Filament voltage (V)	3.5	0	2.6	3.4	0
Lamp current crest factor	1.7	1.7	1.4	1.5	1.5
Ballast factor (%)	93	93	93	100	100
Light output (lm)	5690	5680	5670	5820	582
Flicker (%)	30	30	1	1	1
3rd harmonic (%)	12	12	5	43	43
Ballast efficiency (%)	87	87	89	90	90
Lamp efficacy (lm/W)	77	80	86	93	100
System efficacy (lm/W)	67	70	77	84	90

electronic ballasts for CFL has a power factor above 90%, with harmonics below 20%.

6.3.2 *High-intensity discharge ballasts*

Ballast factors for HID lamps are usually close to unity, and ballast circuitry is somewhat simpler than that for fluorescent ballasts, due to the widespread use of single lamp circuits.

Mercury vapor lamps

Ballast circuits may be of the reactor, autotransformer, or regulator types. Regulator ballasts contain circuitry which is designed to operate the lamp at a relatively constant wattage even though nominal input line voltage may vary. As a rule of thumb, HID ballast losses range between 10 and 20% of lamp wattage.

Metal halide lamps

Most MH lamps operate on a special ballast designed for metal halide lamps. The 1000-W type may be operated on a mercury vapor lamp reactor ballast if the ambient temperature is over 50° F. The metal halide ballast (Figure 6.8) is similar in circuitry to the mercury vapor lamp CWA ballast, but with modifications to provide the higher starting voltage required and wave-shape characteristics to assure reignition of the arc each half-cycle.

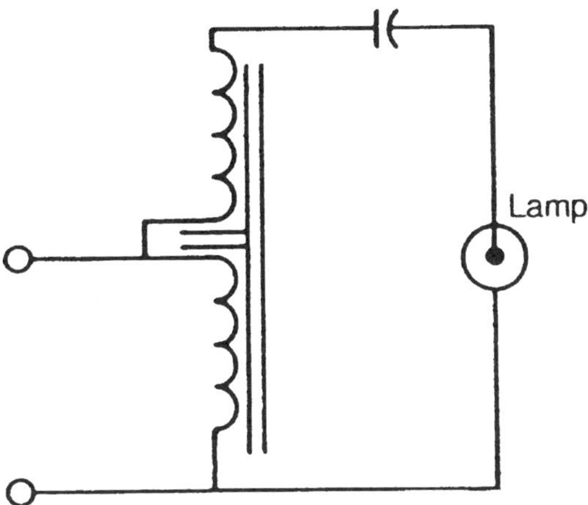

Figure 6.8 Electric circuit diagram for metal halide ballast.

High-pressure sodium lamps

Since no starting electrode is incorporated in HPS lamps, the ballast must supply a high-voltage pulse of 2500 to 4000 V at least once per cycle for wattages other than 1000 W. The 1000-W lamps require 4000 to 6000 V. The element that does this is called a starter or an ignitor. At present, four general types of ballasts are available to the HPS lamp users. Each has its advantages and disadvantages compared with the other in terms of lamp performance, cost, and energy consumption.

1. Reactor or lag ballast — This ballast can maintain good lamp voltage regulation for changes in lamp voltages, but has poor regulation for changes in line voltage. It has a relatively high starting current, which can produce a desirably faster lamp warm-up. It is relatively inexpensive, having low power losses, and is small in size.
2. Lead ballast — This has fairly good regulation for line voltage variations and also for lamp voltage variations.
3. Magnetic-regulated ballast — This is essentially a voltage-regulating isolation transformer with its primary and secondary windings mounted on the same core, and containing a third capacitive winding, which adjusts magnetic flux with changes in either primary or secondary voltage. It provides the best wattage regulation with change of either input voltage or lamp voltage. It has a low-line starting current and a high power factor. It is the most costly and has the greatest wattage loss. Figure 6.9 shows these three types of ballast circuits.
4. Electronic ballasts — The major problem with existing HPS lamp ballasts is their inability to operate the lamps at rated power. Electronic

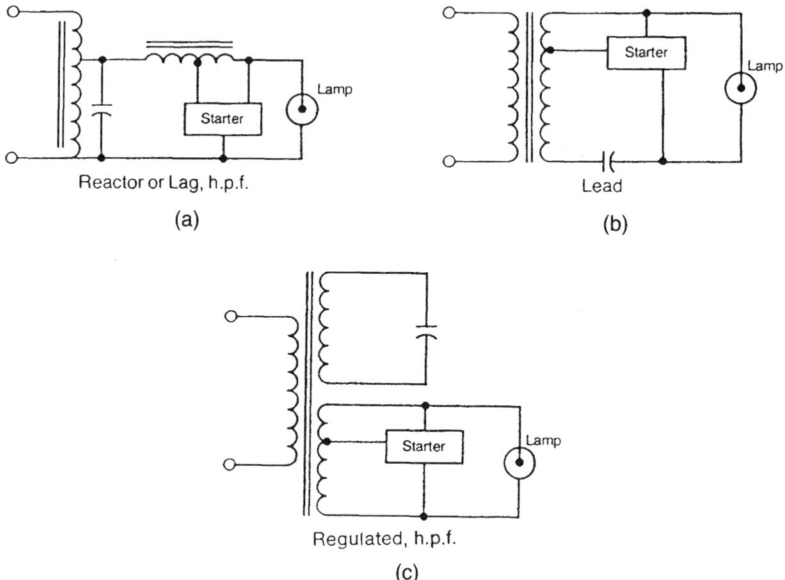

Figure 6.9 Electric circuit diagram for HPS lamp ballast. (a) Reactor ballast circuit; (b) lead-peaked ballast circuit; (c) magnetic-regulated ballast circuit.

ballasts have been built with a solid-state control circuit and a reactor. The use of a solid-state switching device permits the control winding to be shorted in a "phase-controlled" manner, thus providing a smooth and continuous variation in the average inductance of the ballast. The solid-state control circuit monitors lamp and line operating conditions and then establishes the proper value of ballast required to operate the lamp at its rated power. Figure 6.10 shows an electronically controlled HPS ballast. The electronic ballast delivers efficiencies of well over 15% above that of the constant wattage-type core coil–coil ballast. Further, it delivers additional energy savings of 13% by providing constant nominal lamp-rated wattage throughout the lamp life. The electronic ballasts have power factors around 99% and produce line harmonics of approximately 5%.

6.4 Luminaires

6.4.1 Types of industrial luminaires

There are many types of industrial luminaires. Selection of specific types for an installation requires considerations of many factors: candlepower distribution, efficiency shielding and brightness control, mounting height, lumen maintenance characteristics, mechanical construction, environmental

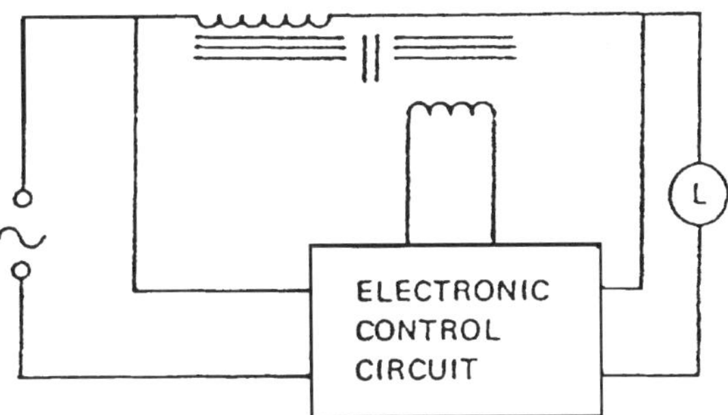

Figure 6.10 Electronically controlled HPS ballast circuit.

suitability for use in normal, hazardous, or special areas. In general, there are five types of luminaires, in accordance with CIE classification for interior applications:

Direct type

Direct-type units emit practically all (90 to 100%) of the light downward to the working area. Although such luminaires usually provide the most efficient illumination on the working surfaces, it is usually at the expense of other factors. For example, shadows may be excessive unless the units have relatively large luminous areas or are mounted closer together than suggested maximum spacing-to-mounting height ratios. But direct and reflected glare may be disturbing because of the higher luminance difference between the bright source and darker surround. Direct industrial lighting equipment is usually classified according to the distribution of the downward component from "highly concentrating" to "widespread". This classification of luminaires is expressed in terms of suggested spacing-to-mounting height ratios, which are shown in Table 6.2. The widespread category includes high-intensity discharge (HID) luminaires that have optical assemblies consisting of a refractor/reflector design that can provide lamp concealment and reduce luminance sufficiently to permit a lower mounting height than would be acceptable for conventional HID luminaires. The distribution of low-bay units tends to improve vertical illumination (because of their wide-angle component) and to permit spacing as much as two or more times their mounting height above the work plane.

Prismatic or mirrored glass or specular aluminum reflectors produce the more concentrating distributions. These are useful when luminaires for general lighting must be mounted at a height equal to or greater than the width of the room, or where high machinery or processing equipment necessitates directional control for efficient illumination between the equipment. They are also useful for supplementary illumination. Spread types are comprised of porcelain-enameled reflectors, other white reflecting

Table 6.2 Classification of Luminaire Direct Component Expressed in Terms of Space Criteria

Spacing-to-mounting height ratio (above work-plane)	Luminaire classification
Up to 0.5	Highly concentrating
0.5 to 0.7	Concentrating
0.7 to 1.0	Medium spread
1.0 to 1.5	Spread
Over 1.5	Widespread

surfaces, diffuse aluminum, mirrored or prismatic glass or plastic, and similar materials. Spread distributions are advantageous in low-bay areas or where there are many vertical or near-vertical seeing tasks.

Generally speaking, concentrating and medium-spread distributions are best suited to high-bay areas. Wherever there is a need for higher-than-average general illumination for an inspection or special work area, highly concentrating luminaires should be installed above cranes at mounting heights where the basic high-bay lighting system is located. For large areas, low-luminance luminaires are preferred to provide low-reflected luminance. Such luminaires may consist of a diffusing panel on a standard type of fluorescent reflector, an indirect lighting hood, or a large luminous area. In a very dusty or corrosive area, luminaires with gasketed glass or plastic covers are recommended.

Area lighting extending from wall to wall is another form of direct lighting in which light from sources in a large cavity of high reflectance is directed downward through cellular louvers or translucent or refracting glass or plastic. When these materials conceal the lamps completely, the illumination characteristics are similar to those of an indirect lighting system. Cellular louvers used as the shielding medium may present a reflected glare problem. This should be minimized in the design. Figure 6.11 shows a typical 400-W high-pressure sodium luminaire that is popular in general factory illumination.

Semi-direct type

These units emit 60 to 90% of the light downward. Utilization of light from these luminaires depends on ceiling reflectance. Light-colored ceilings usually result in improved utilization and visual comfort. The increased ceiling illumination from the semidirect distribution reduces the luminance difference between ceiling and luminaire, increases diffusion, and softens shadows. Approximately designed reflectors or refractors will reduce luminaire luminance and provide additional comfort. Most fluorescent and some HID and incandescent luminaires may be equipped with louvers to further increase shielding and reduce direct glare.

Figure 6.11 A typical 400-W HPS luminaire.

General diffuse or direct-indirect type

In these luminaires, the downward and upward components are approximately equal: 40 to 60% of the total luminaire output. General diffuse-type luminaires emit light about equally in all directions; direct–indirect luminaires emit very little light at angles near the horizontal, which is preferred because of their lower luminance in the direct glare zone. Luminaires with such a distribution are widely used in offices and laboratories, and their use in clean manufacturing areas is increasing.

Semi-indirect type

This type of luminaire emits most of the light (60 to 90%) upward. The major portion of the light reaching the horizontal work plane must be reflected from the ceiling and upper walls; it becomes necessary that these surfaces have high reflectance. The need for high reflectance and good maintenance limits the use of industrial semidirect systems to areas where it is necessary to minimize reflected glare from specular work surfaces.

Indirect type

Indirect luminaires emitting from 90 to 100% of their light upward are seldom used in industry. These units have the lowest utilization and are more difficult to maintain.

Figure 6.12 shows luminaires for general lighting as classified by the CIE in accordance with the percentage of total luminaire output emitted above and below horizontal.

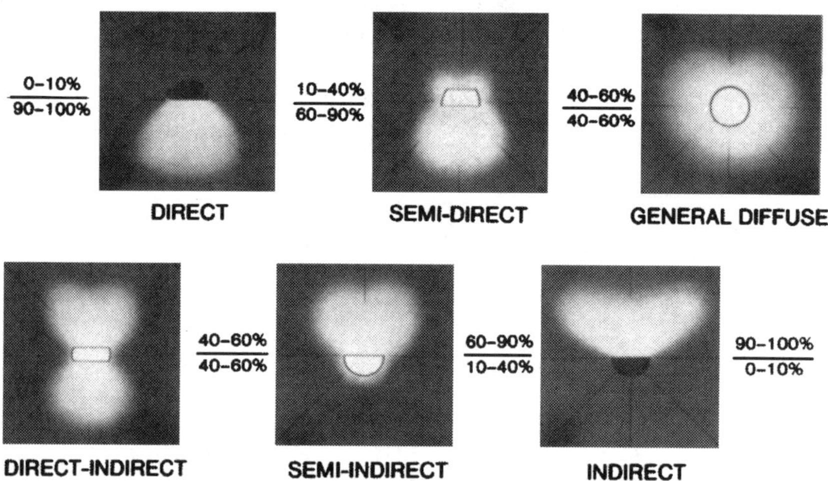

Figure 6.12 General lighting luminaire classifications.

6.4.2 *Supplementary luminaire types*

Supplementary lighting units can be divided into five major types according to candlepower distribution and luminance:

1. Type S-I — Directional: includes all concentrating units, such as a reflector spot lamp or units employing concentrating reflectors or lenses.
2. Type S-II — Spread, high luminance: includes small area sources, such as incandescent or high-intensity discharge. An open-bottom, deep-bowl diffusing reflector with a high-intensity discharge lamp is an example.
3. Type S-III — Spread, moderate luminance: includes all fluorescent units having a variation in luminance greater than 2:1.
4. Type S-IV — Uniform luminance: includes all units having less than 2:1 variation of luminance. Usually, this luminance is less than 6800 cd/m^2 (2000 fL). An example of this type is an arrangement of lamps behind a diffusing panel.
5. Type S-V — Uniform luminance with pattern: a luminaire similar to type S-IV, except that a pattern of stripes or lines is superimposed.

6.4.3 Specular reflectors

In recent years, specular reflectors have been promoted as a potential source of energy savings for fluorescent lighting systems. A specular reflector is a luminaire component that has a highly polished surface. Applications of specular reflectors can increase luminaire efficiency, thus reducing the number of lamps, ballasts, and/or luminaires that would be required in the system.

Specular reflectors can be used in new fluorescent luminaires or installed in existing luminaires as a retrofit strategy. In general, specular reflectors are made of one or more of three material types: anodized aluminum, enhanced anodized aluminum, and silver film that is applied to a metal substrate. The long-term performance of the retrofitted luminaire is affected by its material's characteristics. Any degradation of material during its life will affect the luminaire's ongoing performance. Figure 6.13 shows the position of a specular reflector within a luminaire. The existing lamps are replaced with new lamps, and the luminaire and lens are cleaned. When these steps are taken, the connected power to the luminaire is approximately half of that prior to the retrofit, but the average illuminance may be greater than half by a substantial margin.

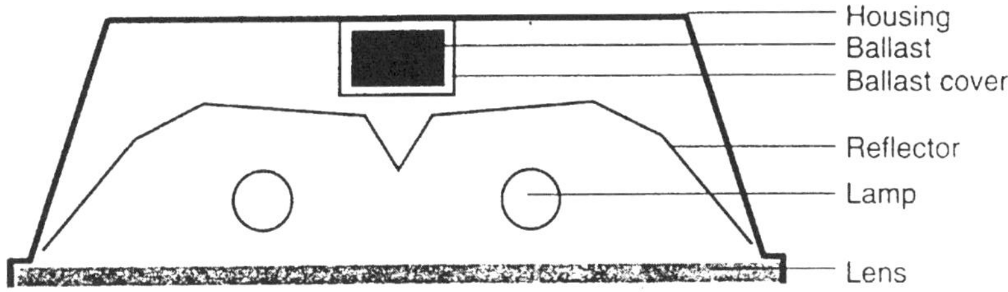

Figure 6.13 The position of a specular reflector within a luminaire.

Also, a recent design uses T8 lamps and electronic ballasts in a low-profile housing with lenses and special reflectors to achieve high luminaire efficiency and a broad distribution, while allowing mounting close to the ceiling.

References

Chen, K., *Energy Effective Industrial Illuminating Systems,* The Fairmont Press, Lilburn, GA, 1994.

Knisley, J. R., Updating Light Sources for New and Existing Facilities, EC&M, November, 1990.

NLPIP Specifier Reports on Electronic Ballasts, Rensselaer Lighting Research Center, Troy, NY, December, 1991.

NLPIP Specifier Reports on Specular Reflectors, Rensselaer Lighting Research Center, Troy, NY., July, 1992.

NLPIP Specifier Reports on Compact Fluorescent Lamp Products, Rensselaer Lighting Research Center, Troy, NY, April, 1993.

Verderber, R.R., Morse, O., and Rubinstein, F.M., Performance of Electronic Ballast and Controls with 34 & 40 W F40 Fluorescent Lamps, in Proc. IEEE-IAS Conf., San Diego, CA, October, 1989.

chapter seven

Energy-oriented new and retrofitting installations

7.1 Introduction of the new trend

Several forces have shaped how the use of commercial lighting equipment has evolved over the last two decades:

1. An increasing emphasis on obtaining higher worker productivity — An interest in quality lighting that improves the employee's visual task performance even by a few percent can quickly pay for itself through increased productivity.
2. Moving toward an electronic office environment — The advent of the visual display terminal (VDT) raised concern for the worker's visual comfort. At the same time the introduction of easily changeable, open-plan office furnishings, including cubicle walls of varying heights plus overhead file and shelves, has necessitated the installation of additional luminaires to offset shadows caused by those structures. This additional light on the work task has produced an opportunity to reduce the quantity of ambient light from the general lighting system. It also led to the need to improve the quality of the system's available light in order to reduce glare on VDT screens.
3. The lighting equipment market is legislated — In the late 1980s lawmakers established minimum efficiency standards for ballasts that operate the fluorescent lamps in the most commonly used luminaires made since April 1991. At the same time, the Energy Policy Act of 1992 is affecting lamps. First in May 1994 with the most popular 8-ft lamps, then in November 1995 with the most popular 4-ft straight and 2-ft U-shaped lamps, this law prohibited the least efficient models from being produced or imported in the U.S. It also banned the most inefficient incandescent reflector lamps. The net effect of these laws will be to raise minimum efficiency levels for both major components of fluorescent luminaires. Another result of the Energy Policy Act is

the federally mandated requirement that each state adopt laws to meet minimum energy efficiency standards for both new construction and renovations. For lighting, these standards can take the form of maximum unit power density limits in watts per square foot for each space within and outside a facility. In some cases the use of lighting controls — photocells, time clocks, occupancy sensors, and/or ambient light level sensing plus dimming — will be necessary to achieve these standards.

4. Environmental issues — PCBs were banned from ballasts in 1979, followed in 1991 by elimination of DEHP. Both were used in the p.f. correction capacitors of most ballasts. More recently manufacturers have begun to reduce the amount of mercury in each lamp to a level much lower than before. Disposal of lighting equipment containing those materials is regulated under one or more federal laws. Both fluorescent and HID lamps as well as types of ballasts containing toxic materials can now be recycled by qualified and licensed waste disposal organizations.
5. Proper maintenance — It will be much more cost effective to periodically clean the lens and inside of the luminaire. This type of proactive lighting maintenance can result in energy saving with improved quality of lighting; reduced lamp and ballast inventory carrying costs; and reduction of unnecessary ballast degradation from trying to start expired lamps.

7.2 Examples of new installations

7.2.1 Machine shops

Machining of metal parts consists of setting up and operating machines such as lathes, grinders, millers, shapers, and drill presses, bench work, and inspection of metal surfaces. The precision of such machine operation usually depends on the accuracy of the setup and careful use of the graduated feed-indicating dials rather than observation of the cutting tool or its path. The fundamental seeing problem is the discrimination of detail on plane or curved metallic surfaces.

The visibility of scribed marks depends on the characteristics of the surface, the orientation of the scribed mark, and the nature of the light source. Directional light produces good visibility of scribed marks on untreated cold-rolled steel if the marks are oriented for maximum visibility, such that the brightness of the source is reflected from the side of the scribed mark to the observer's eye. Unfortunately, this technique reduces the visibility of other scribed marks. Better average results are obtained with a large-area, low-luminance source.

There is an obvious advantage in the use of large-area, low-luminance sources for most visual tasks in the machining of metal parts. The ideal general lighting system is one having a large indirect component. Both fluorescent and HID sources can be used for general lighting; fluorescent luminaires in a grid pattern are usually preferred. High-reflectance room surfaces improve visual performance. Figure 7.1 shows the result of such a system, which provides a pleasant environment for die making in a machine shop. Supplementary lighting is used to maintain close tolerances of die production. Green plants and modern wall treatment contribute to making a stimulating space.

Figure 7.1 Fluorescent lighting for a machine shop with supplementary lighting for die production.

Special types of luminaires are designed to illuminate three-dimensional objects such as machinery, dials, spindles, presses, panels, and stacked materials. Light is contributed from many locations and distances to minimize shadow, reduce glare, and optimize visual comfort. This type of luminaire has proven to be satisfactory for lighting of machine shops. Figure 7.2 shows a different type of lighting for another industrial plant machine shop, which is lighted with 250-W metal halide luminaires with emphasis on vertical surface illumination.

7.2.2 Control rooms

The control room is the nerve center of a power plant or process plant and must be monitored continuously. Lighting must be designed with special attention to the comfort of the operator; direct or reflected glare and veiling

Figure 7.2 Metal halide lighting for an industrial plant machine shop.

reflections must be minimized, and luminance ratio must be low. Along with ordinary office-type seeing tasks, it is often necessary to read meters 10 or 15 ft away.

Although the practice is not standardized, most control room lighting involves one of two general categories: diffuse lighting or directional lighting. Diffuse lighting may be from low-luminance, luminous indirect lighting equipment, solid luminous plastic ceilings, or louvered ceilings. Directional lighting may be from recessed troffers that follow the general contour of the control room.

A basic conflict exists in trying to light a control room, since some tasks are made more visible under reduced illumination, while other tasks require significantly higher levels of illumination. Fortunately, most of the tasks enhanced by low light levels are not located in the same area where tasks requiring high illumination are to be found. Therefore, the lighting system can be modified to give nonuniform distribution of light within the room.

The optimum distribution of illumination for each control room must be determined on a case-by-case basis since the equipment arrangement will vary in each installation. It may also be found necessary to use several different types of louvers, lenses, and diffusers, sometimes in combination with one another, to achieve the optimum distribution for each particular room.

7.2.3 VDT rooms

Currently, there is a proliferation of VDTs in all areas of industry and commercial establishment. Lighting for these areas needs special attention. The lighting design should provide for a good visual performance, comfort, and creation of a psychologically and aesthetically pleasing environment. In specific terms, the lighting design should still limit both direct and indirect glare and control luminance in both the immediate task area and within the dynamic field of view.

Consider the following:

1. Illuminance — Illuminance values can be determined from the *IES Handbook*. Most VDTs are self-luminous and require virtually no illuminance for task contrast. However, one must consider adjacent tasks, comfort, and psychological well-being. As many adjacent tasks will fall into category "D", illuminance values of 20–30–50 fc are most likely to be appropriate. Typically, 20 to 40 fc of general illumination should be the maximum, with task lighting supplementing where necessary, with quality (no veiling reflection), controlled brightness luminaires.
2. Luminance — Luminance is illuminance times reflectivity. Current practice suggests that luminance ratios are important to good visual performance and comfort. Within the "task surround", luminance values should not be greater than three times that of the task, or less than 1/3 the level of the task. These standards may be conservative but provide a useful guideline. In measuring a variety of VDTs, screen luminance typically ranges from 5 to 25 fL, with 15 fL a reasonable average.
3. Luminaires and indirect glare — The most common complaint of office lighting today is indirect glare on VDT screens. All types of indirect glare are a result of brightness in the "offending zone" (see Figure 4.8).
4. Light fixture location — The best place to locate fixtures is to the side of the VDT; such positioning will minimize reflection of the fixture on the screen. Figure 7.3 shows such an arrangement. Optimum placement of the fixture for a particular VDT might cause problems for other VDT operators. So conscientious planning is required for both the positioning of light fixtures and the placement of work stations. The use of a three-lamp fixture should be considered because such luminaires are more energy-efficient, and they permit three levels of switching. Another basic method to eliminate fixture reflections in the face of the VDTs is to equip the fixture with a specular parabolic wedge louver, with an absolute cutoff of 45°, so lamp images will be reflected below the viewing angle.

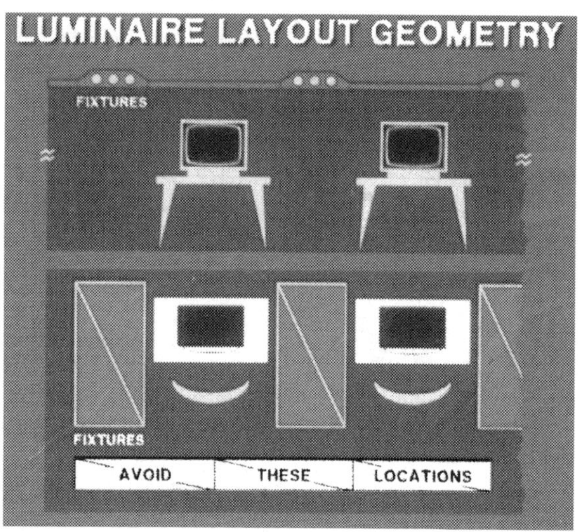

Figure 7.3 Well-shielded luminaires placed to the sides of the work station.

7.2.4 Manufacturing areas

Fluorescent lighting was widely used for industrial operations prior to the energy crunch in the mid-1970s. During the last decades, HID (particularly the metal halide and high-pressure sodium) lighting has become more applicable in industrial plants.

In high-bay areas, 1000-W HPS luminaires, which provide some uplight component through open and ventilated reflectors, are used in an approximate spacing-to-mounting height of 1:1. For low-bay areas, 400-W luminaires with about a 16-ft mounting height provide even candlepower distribution of light without glare using a faceted aluminum reflector and a polycarbonate lens having refractor prism elements. The optical assembly is totally enclosed, gasketed, and filtered to keep contaminants from infiltrating to the lamp and reflector or inside the lens. Since the lens prisms redirect some of the high-angle lumen output toward lower angles, these industrial units provide adequate horizontal beam spread and vertical surface illumination.

In general, the spacing and mounting-height arrangements of both low- and high-bay units provide approximately 60 fc horizontal maintained on a typical work surface. All the fixtures are wired in a checkerboard pattern on 480-V three-phase circuits. The overall light level can be reduced 50% in a given zone by switching off half the circuits in that zone from a convenient central point. Thus energy savings can be achieved during reduced production activity. Each fixture is powered through a fused plug and cord assembly which permits easy replacement of the unit for repair.

7.2.5 Warehouses

Designing a lighting system for a warehouse might appear to be relatively simple. Install enough luminaires to deliver enough light onto the stored material to permit accurate selection, and in the area for operators to see safely where they are going. Over the years, a rule of thumb has evolved as an offshoot of manufacturing area lighting practice. However, in recent years the concept of storage has changed. Warehouses are larger, stacks are higher and deeper, and operations have become more automatic and computer controlled. In designing lighting for most new warehouses, illuminating engineers find their job comparable to designing a system for a long, tall, narrow room for which reflections are unknown and in which luminaires can be placed only in high, inaccessible places.

In the past (before energy crunch years), fluorescent lighting was commonly used to illuminate vertical surfaces. Figure 7.4 shows high-output fluorescent luminaires operating on 277-V circuits installed at a mounting height of 40 ft in a modern warehouse. It delivered satisfactory vertical illumination, especially near the top of stacks.

Today, higher mounting heights have prompted the use of HID lighting equipment. Because the light is emitted from an optically small area, a HID lamp permits finer optical control than do line source fluorescent lamps; luminaires can be designed with better directional characteristics than those for line source.

In general, several factors are unique in warehouse lighting. First, there is the necessity of seeing on a vertical surface rather than on a horizontal plane. Second, the warehouse aisle is similar to a narrow, long room with high walls, where interreflections affect the result to a significant degree. Third, the type and amount of material in the warehouse aisle are subject to unpredictable fluctuation.

Recent studies on warehouse aisle lighting have pointed out that luminaires with maximum uplight provide superior vertical as well as horizontal illumination at all locations for several aisle widths. The uplight would be reflected from the ceiling, which would be advantageous in a warehouse aisle setting with narrow aisles and high mounting heights. The study concluded that the amount of uplight appears to be an important factor and that luminaire design appears to be one of the most critical elements in the production of vertical illumination. Figure 7.5 shows an empty warehouse lighted with 400-W HPS luminaires, resulting in good vertical illumination. The HPS lamps were also chosen because of the cost factor.

Illuminating Engineering Society (IES) has published a "Design Guide for Warehouse Lighting". Table 7.1 lists recommended maintained illuminance values for the various storage, shipping, and receiving spaces in a warehouse. Illuminance ranges in footcandles or lux are to be used for design purposes as targeted maintained values. Proper use of this table in conjunction with Tables 4.3 and 4.4 can assist illuminating engineers to determine a most appropriate illuminance value required for the design.

Figure 7.4 High-output fluorescent lighting for warehouse aisles.

7.2.6 Plant engineering offices, conference rooms, and hospital rooms

Engineering offices — Visual requirements for engineering and drafting offices demand high-quality illumination since discrimination of fine detail is frequently needed for extended periods. Harsh directional shadows from drawing instruments may reduce efficiency. Illumination systems that avoid reflections are most important in providing maximum contrast. Studies indicate that ideal locations for lighting fixtures were at least 24 in. from either side of the drafting table. In general, the recessed 2×4-ft parabolic fluorescent troffers with 4-in.-deep cells and four lamps are suited for such applications. This type of lighting system can produce virtually glare-free and adequate illumination at the center of the drafting tables.

Figure 7.5 400-W HPS lighting for an unoccupied warehouse.

Conference rooms — The diversity of work to be performed in the conference room requires that the illumination should be flexible and the entire room should be comfortable and pleasant. The general lighting is provided by the recessed fluorescent in the 2 × 4 grid-type ceiling. Specifically chosen reflectors provide high-level illumination without glare. Incandescent (250-A PAR 38) downlights with concentric louvers are provided in the front

Table 7.1 Recommended Illuminance Categories for Warehouse Lighting

Type of activity	Illuminance category	Lux	FC
Inactive or rough and bulky storage	B	50–75–100	5–7.5–10
Active or medium storage	C	100–150–200	10–15–20
Small item storage	D	200–300–500	20–30–50
Shipping and receiving	D	200–300–500	20–30–50
TV surveillance	Contact manufacturer of TV camera		

Notes: Recommended footcandles are based on maintained average illuminance levels and should include the proper total light loss factor.

Recommended footcandles are for both vertical and horizontal surfaces.

Uniformity of illumination measured in terms of average minimum should not exceed 10:1 on vertical surfaces and 3:1 on horizontal surfaces.

stage and over the conference table. This, in combination with dimmer-controlled general lighting, provides flexibility to create a changing environment, as well as eye comfort for all seeing tasks. Figure 7.6 shows a handsomely decorated functional conference room.

Figure 7.6 Combination of fluorescent and incandescent lighting for a functional engineering conference room.

Hospital rooms — A plant hospital room is used for emergency and/or simple treatments only. It is staffed with a nurse and a part-time doctor. In the hospital room a moderately high illuminance is required for examination. Ideally, a 50-fc level for local examination is recommended. It is advantageous to use three- or four-lamp fluorescent luminaires for the general illumination, switched or dimmed to allow reduction of light level during certain procedures. This type of room can be better served with general illumination supplemented by a portable of fixed examination light. The general light source should be of the color-improved type. Figure 7.7 illustrates such a design.

Figure 7.7 Level-controlled fluorescent lighting for a plant hospital room.

7.3 Retrofitting installations

7.3.1 The significance of energy efficiency in retrofitting

It has been recognized that energy-effective illumination design has a tremendous impact on the costs. Here the costs include not only the initial cost of the installation, but also the operating and maintenance costs. Often energy-effective design will result in an increase of the initial cost over the traditional approach. However, often when lighting quality is improved as a result of successful retrofitting, higher worker productivity as well as energy/cost savings can be achieved. Therefore, the engineer or designer must not fail to recognize that savings can result from improved productivity, few errors and rejects, increased safety and security, lower liability insurance premiums, increased retail sales, etc. In many instances, the value derived from better design and hence the better performance of the retrofitting system is worth ten or more times what would be realized from even the most significant reduction in energy consumption. In short, the engineer or designer must not lose sight of the importance of the system performance to the energy savings in the initial stage of the project.

7.3.2 Examples of retrofitting installations

The following chosen examples are intended to illustrate several excellent projects which have achieved not only energy/cost savings, but improved employee's visual performance as well. One example is purposely included here to illustrate a total failure due to a nonprofessional's recommendation, which treated the retrofit project as a replacing game focusing on the energy savings alone without studying the worker's visual requirements, and resulted in an unsalvageable installation.

Retrofit options for a mercury lighting system

The high operating cost of a mercury lighting system is second only to that of an incandescent system. For maximum savings, a mercury system can be replaced with a high-pressure sodium (HPS) system, or if color discrimination is important, a metal-halide (MH) system. Two basic retrofit options can be applied to a mercury system. The first option is to replace the luminaire ballasts with either MH or HPS units. The second option is to replace with special MH or HPS lamps designed to operate on mercury ballasts. Replacing ballasts requires a larger cash expenditure, but the long-term benefits are greater. Although retrofit lamps result in substantial savings, they generally cost more and have a shorter life, or lower efficacy or lumen maintenance than those of standard MH or HPS lamps operated on companion ballasts.

At the present time, retrofit HPS lamps in wattages of 150, 215, and 360 W can be used on mercury systems using 175-, 250-, and 400-W lamps, respectively. One manufacturer also offers 880-W HPS lamps for use in existing 1000-W mercury luminaire. Retrofit MH lamps are available in 400- and 1000-W versions, which are intended for use with CW/CWA-type mercury ballasts. The 1000-W MH can reduce wattage per luminaire by either 35 or 85 W, depending on the type of ballast it is used with. A new addition is the 325-W MH lamp which saves about 70 W per luminaire while delivering 40% more light than the 400-W mercury lamp it replaces.

Fixture for fixture replacement: This pattern of relighting represents one of the simplest conditions. A typical case for such a system is to replace 1000-W mercury lights with 400-W HPS on a one-for-one basis. Usually, the spacing-to-mounting height ratio is suitable for the new light source. The existing wire sizes are usually adequate and connections can easily be made. The overall installation costs of the retrofit program will be low and the ROI will be high. The lighting level will stay practically the same as before. The energy savings realized in this retrofit amounts to 60% of the original consumption. Figure 7.8 shows a typical high-bay plant relighted in this nature. On the left, 178 1000-W mercury luminaires were used to illuminate the plant; on the right, the same number of 400-W HPS new luminaires are in place. The ROI in this case is 58%.

Figure 7.8 A typical industrial high-bay retrofit project – 400-W HPS replacing 1000-W mercury.

Federal building and courthouse in San Diego

A lighting retrofit at the Federal Building and Courthouse which is expected to cut the facility's annual energy consumption by 2.3 million kWh was encouraged by a utility rebate covering over 20% of the project cost. The total cost including change orders and lamp and ballast recycling was $1,333,000. Local utility awarded a $262,000 rebate for the retrofit. The expected annual savings will be $230,000. The payback is about 4.7 years.

The project was completed in September 1994. It involved replacing lamps and ballasts in 10,500 office fixtures. Reflectors, photocells, and motion detectors were installed. As a result of the lighting upgrades, the building's electrical demand dropped by 1380 kW, which is made up by:

1. 450 kW reduction from lamp and ballast replacement
2. 850 kW reduction attributable to the motion sensors
3. 80 kW reduction from daylighting with photocells

Throughout the project, 21,660 32-W T8 lamps were used to replace 40-W T12 fluorescents, and a total of 14,000 electronic ballasts were installed to replace magnetic units. Rather than install a separate lighting control system, the project was opted to utilize motion detectors and photocells for lighting control. All sensors were wired directly into the lighting circuits. No computerized control was used. This kept control equipment, installation, and maintenance costs down while still providing energy savings.

New post office building in California

The building was not designed to comply with Title 24 because it was a federal building. Negotiating with the utility the building was able to secure a rebate for daylighting under the utility's Title 24 program, which pays for energy efficiency measures that beat the state building code by at least 1%.

The daylighting system's skylights, roof closures, and controls cost $540,000 and are expected to cut 3,334,000 kWh in energy consumption annually. This is expected to save $147,000 from the facility's utility bills. With a rebate, the payback for daylighting will be about 2.7 years.

The daylighting system is used to supply lighting for a single-story workroom. The building's remaining area over two floors does not include daylighting controls for its electronically ballasted T8 lighting.

The workroom daylighting system includes 280 skylights covering 3.4% of the workroom roof. Each skylight measures about 4 × 8 ft and includes an acrylic lens with ultraviolet light stabilizers. They have a viable light transmittance of 72% and a shading coefficient of 66%. The system also includes a master time control to control 1650–250-W lamps. The luminaires are configured in a diamond pattern in each of seven zones.

Based on photocell readings, the controller can turn off one to three lamps in a diamond when ambient light is sufficient to meet set points. The rest of the building lighting consists of 32-W T8 fluorescent lamps with electronic ballasts. The installation of 9782 T8 lamps and 3820 two-, three-, and four-lamp ballasts is expected to cut an additional 515,000 kWh from the building's annual electric energy consumption.

Greening the White House

The challenge is fascinatingly complex, due to the many functions the building serves. The retrofit measures must be carried out without disturbing the residents, and must be essentially invisible to the visitors, although the work is projected over several phases and will take several years to complete. Here the discussions are limited to the lighting only. Short-term lighting actions focus on improving energy efficiency and management of existing lighting while increasing daylighting where possible. One guiding principle is the increasing of task lighting for computer and other desk work, while reducing overall ambiance to save energy. The following are some specific design modification ideas:

1. Replacing table lamp of incandescent bulbs with compact fluorescent where appropriate. Typical energy savings are 65 to 75% per lamp. This saves more than $100 over the lifetime of each lamp and yields a 6-month payback.

2. Accelerating the West Wing lighting upgrade, which cuts energy use 50 to 70% and yields a 3-year payback. This upgrade is based on replacing T12 fixtures with magnetic ballasts with T8 and electronic ballasts.
3. Upgrading concealed bathroom incandescents with compact fluorescents. This will cut energy use by 65 to 75%.
4. Upgrading the fluorescents in residence service areas with T8 lamps, electronic ballasts, and occupancy sensors. The investment will yield a 20% energy savings, and the estimated payback is 5 years.
5. Replacing exterior flood with energy-efficient lighting will result in energy savings of 40%.
6. Replacing T12 fixtures with T8 lamps and electronic ballasts in the Old Executive Office Building.
7. Improving the efficiency and color rendering of the down lights in the Indian Treaty Room. Mercury vapor lights have been replaced with metal halide lamps that offer improved color rendition and 40% more light at no additional cost.
8. Rehabilitating existing historic skylights in the hallways on the fifth floor of the Old Executive Office Building to increase available daylight. The existing skylights have been dirty and prevent decent light transmission.

In general, implementation of the above actions will improve comfort and productivity and will reduce energy use from 20 to 75% per each action.

A telephone company's project

This project involved replacing four 40-W T12 lamps and two magnetic ballasts in each of the 1600 fixtures with two 32-W T8 lamps, a single electronic ballast, and a reflector. This retrofit reduced demand from 192 to 62 W per fixture to a total demand reduction of approximately 208 kW. Annual energy consumption consequently fell by 865,000 kWh, based on 4160 h of operation per year. The installation will cut operating costs at the 163,000-ft^2 building by an estimated $63,000 a year. A projected $52,000 in annual utility bill reduction is based on average savings of $4,300 realized between February and June. An additional $5,600 in maintenance and equipment cost savings comes from reduced ballast failure and fewer lamp failures. Finally, $5,700 savings comes from the estimated 41-ton reduction in cooling demand stemming from the delamping.

The project's 1.7-year simple payback is further shortened by a $41,000 incentive paid by the utility. This incentive brought the project's payback to just under 13 months.

A machine shop

This example illustrates the consequence of playing a replacing game to save energy while ignoring the recommended practice in lighting design.

The existing machine shop has a 160 × 200-ft production area in which medium to fine machining work is performed. The existing system consists of stem-mounted 8-ft high-output (HO) fluorescent fixtures in continuous rows on 12-ft centers as shown in Figure 7.9. These fixtures provided a uniform horizontal maintained illuminance of 70 fc. The plant managers were ill advised by a nonprofessional consultant that lighting costs could be reduced by 50% through the use of HPS lighting. Specifically, individual 400-W HPS fixtures stem mounted at 25 ft above the floor were installed. Immediately after, employees began complaining about poor visibility at their work stations.

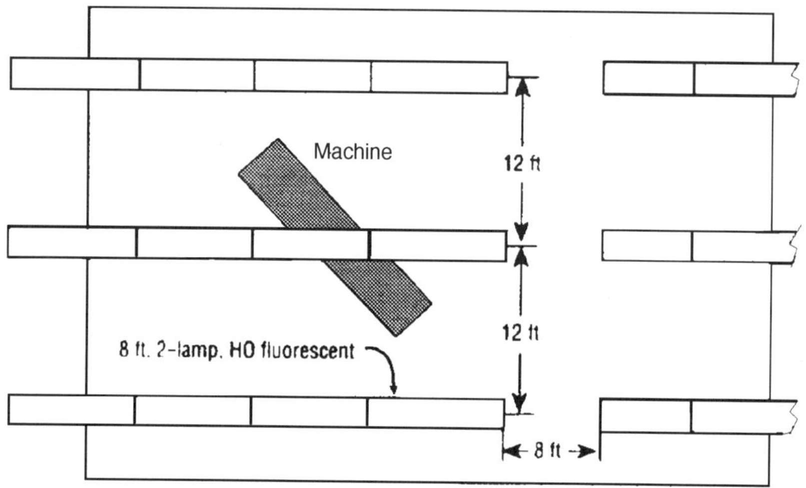

Figure 7.9 A machine shop layout with fluorescent lighting.

After analyzing the data obtained from the site, two important design practices were found to be violated. First, the designed illuminance level was far below that recommended by IES for the type of work being performed. Second, the photometric distribution pattern of the luminaire was not compatible with the fixture layout within the area. The first error was supported by the light meter readings, which ranges from 10 to 40 fc on the horizontal work plane, far below the 70-fc level of the original fluorescent lighting and that called for by IES. The second error was established after studying the HPS fixture manfacturer's data sheet. In this case, the HPS luminaire has a hemispherical reflector and a prismatic refractor lens enclosing the lower part of the reflector's housing. The distribution of this fixture concentrates the greatest amount of light within a 25° angle on either side of the nadir. This fixture is termed a sharp-cutoff unit. Consequently, most of the lumen output falls only within 20 ft diameter circular area of the floor

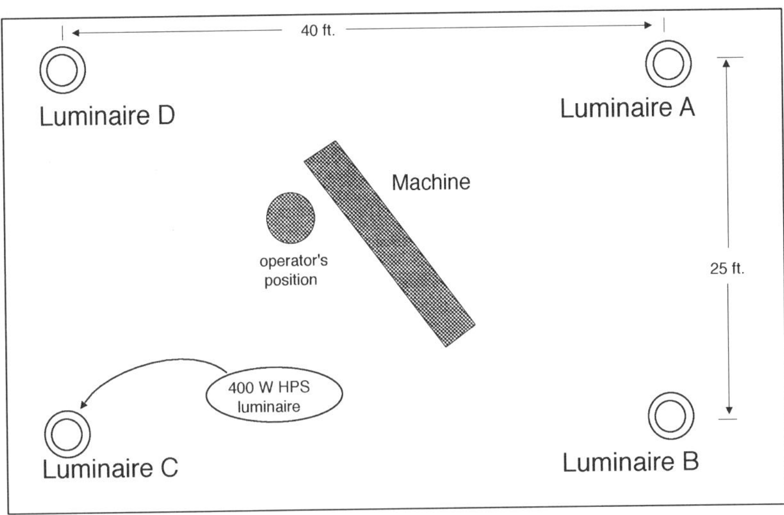

Figure 7.10 A typical four-fixture work station layout.

for this mounting height, and directly below the fixture. An ideal application for this type of fixture would require a mounting height of 40 ft or more.

An even more revealing error was uncovered by carrying out a point-by-point computation. Figure 7.10 shows a four-fixture work station layout, typical of many of the work stations. Table 7.2 shows the calculated light contributions of each of the four fixtures. Note that fixture A contributes over half of the light. Problem areas surface in the form of shadows created by the operator when blocking light from fixtures A and D. Machine positioning also creates problems because fixtures B and C contribute relatively little light to the task area.

The light distribution of a suggested replacement HPS luminaire indicates that it would contribute light at larger angles from the nadir and would be better suited to the 25-ft mounting height application. Should this type of luminaire be specified, all four of the fixtures would have a

Table 7.2 Light Contributions from HPS Fixtures Originally Installed

Luminaire	Angle from nadir	Candlepower (cd)	Horizontal footcandles	Contribution (%)
A	17	16,000	18.79	53
B	39	6,800	4.29	12
C	40	6,800	4.11	12
D	39	10,500	8.32	23
		Totals	35.51	100

more even share of light contribution, as given in Table 7.3. Also note that the illuminance on the task is increased by 20%. However, the average maintained illuminance level would still be below the recommended 70 fc. So the final solution is to reinstall the original fluorescent HO luminaires, but retrofit them with electronic ballasts to obtain 25% reduction in energy costs. In addition, this system will produce fewer flickers as reported in its operation.

Table 7.3 Light Contributions from Suggested HPS Fixtures to be Mounted in Same Locations

Luminaire	Angle from nadir	Candlepower (cd)	Horizontal footcandles	Contribution (%)
A	17	10,200	11.98	29
B	39	14,000	8.83	21
C	40	14,100	8.51	20
D	33	16,200	12.83	30
		Totals	42.15	100

References

Chen, K., Design Applications in Lighting Retrofits, IEEE-IAS Annu. Conf. Proc., October, 1992.

Ferzacca, N. D., Avoid the Pitfalls of Lighting Retrofit, EC&M, pp.59–62, April, 1993,

Gahran, A., GAS facility uses utility rebate to leverage cost of Lighting Retrofit, *Energy User News*, February, 1994.

IES, Design Guide for Warehouse Lighting, IES DG-2-1992, Illuminating Engineering Society of North America, New York.

Rowe, G. D., Solving lighting problem in VDT areas, *Plant Eng.*, December 27, 1984.

Watson, C. M., Fixture components critical in retrofit decisions, *Energy User News*, May, 1996.

chapter eight

Evaluating and achieving energy efficiency

8.1 Achieving energy effectiveness in a new system

8.1.1 General requirements for industrial visual tasks

Industry encompasses seeing tasks, operating conditions, and economic considerations of a wide range. Visual tasks may be extremely small or very large; dark or light; opaque, transparent, or translucent; on specular or diffuse surfaces; and may involve flat or contoured shapes. With each of the various task conditions, lighting must be suitable for adequate visibility in developing raw materials into finished products. The speed of operations may be such as to allow only minimum time for visual perception, and therefore lighting must be a compensating factor to increase the speed of vision.

The lighting system should be a part of an overall planned environment. The design of a lighting system and selection of equipment may be influenced by many economic and energy-related factors. For related information, reference should be made to Chapters 4 and 6.

8.1.2 Quantity and quality of illumination

lluminance recommendations for industrial tasks and areas are given in Tables 4.1 and 4.2. However, illuminance values for specific operations can be determined using illuminance categories of similar tasks and activities found in those tables, and the application of the appropriate weighting factors in Tables 4.3 and 4.4. In either case, the values given are considered to be target-maintained illuminance. To ensure that a given illuminance will be maintained, it is necessary to design a system to give initially more light than the target value.

The quality of illumination pertains to the distribution of illuminance in the visual environment. Glare, diffusion, direction, uniformity, color, luminance, and luminance ratios all have a significant effect on visibility and the ability to see easily and quickly.

Industrial installations of very poor quality are easily recognized as uncomfortable and are possibly hazardous. The cumulative effect of even slightly glaring conditions can result in material loss of seeing efficiency and undue fatigue.

8.1.3 Design considerations

The illuminating engineers who are to design a system for industrial environment should consider the following as the important requirements:

1. Determine the quantity and quality of illumination desirable for the manufacturing processes involved.
2. Select lighting equipment that will provide these requirements by examining photometric characteristics and mechanical performance that will meet installation, operating, and actual maintenance conditions.
3. Select and arrange equipment so that it will be easy and practical to maintain.
4. Apply control and daylighting techniques to the proposed system for promoting overall energy effectiveness.
5. Balance all of the energy management considerations and economic factors including initial, operating, and maintenance costs vs. the quantity and quality requirements for optimum visual performance.

8.2 Achieving energy effectiveness in a retrofit system

When specifying a retrofit system, the goal is to maximize energy efficiency in addition to meeting the necessary lighting requirements for the occupants in the area. The system approach requires a thorough review of all aspects of the lighting system to be retrofitted. Each area of the facility should be studied to determine the task performed. An examination should be conducted of the existing lighting components that currently provide the illumination for the tasks. Once the survey has been completed and evaluated, an appropriate retrofit system can be specified for each area. Proper design based on the steps described in Chapter 4 should follow. To make sure that the proposed system satisfies the objectives of energy effectiveness, the illuminating engineers must also not forget the application of the control and daylighting techniques to the system wherever feasible. The necessary steps to be carried out for the retrofit projects are the following.

The survey — Accurate information is needed on what exists and how it is being used. Survey forms are useful and data can be collected on a hand-held dictating machine and transcribed to a spreadsheet or a database for analysis.

Each system must be noted as to its type, designated by the name of the representative fixture for each system, the type of lens or louver used, the method of mounting, the lamp type, the number of lamps per fixture, the type of ballast and number per fixture, the type control for each fixture subsystem, the tasks performed in the area, and the type of ceiling. To determine whether or not certain fixtures can be tandem wired if multilamp electronic ballasts are to be used, the illuminating engineer should find out if there are any special occupancy use requirements, the age of the occupants, and whether computers or other video display terminals are used in the space. Levels of ambient illumination can be reduced if the primary work is on a computer screen or suitable task lighting may be applied.

Specify the retrofit — Once all the above data have been collected, the most efficient way to optimize the lighting system in each area should be determined. The proposed retrofit should minimize the energy input required to produce the appropriate illumination. In carrying out a lighting retrofit project, some of the key functional and technical performance factors as listed below should be carefully considered:

Functional performance	Technical performance
Task requirements	Lamp lumen depreciation
Illuminance values	Efficacy or efficiency
Color rendition	Energy density
Daylighting	Life cycle cost
Aesthetics	Maintenance
Controls	Power factor
Glare	Harmonics

It must be remembered that (1) lighting levels will decrease or diminish with time and (2) a total fixture upgrade is a more effective approach. The use of occupancy sensors should only be considered for the right application. Task lighting in the work area should also be considered as this may allow for a substantial reduction in ambient lighting. Another area to investigate is daylighting, which can be used to supplement existing lighting systems and has become an effective means of saving energy.

Implementation — When retrofit system design is complete, the next step will be to focus on buying and installing the required components. Separately bidding the materials and the installation usually offers project cost savings. Special care should be taken to specify responsibility for disposal of PCB-containing ballasts and mercury-containing fluorescent and HID lamps. In addition to the materials and installation bids, engineers in charge should also contact the local utility to determine the extent of its financial assistance to the lighting retrofit. Some utilities negotiate rebates. Each utility may have a slightly different program for rebate.

8.3 Evaluating energy effectiveness in a new or retrofit system

8.3.1 Auditing

The audit report should contain the following information:

1. Existing data — should consist of the following:
   - Number and wattage of existing fixtures
   - Demand in kW attributed to lighting
   - Burning hours
   - kWh usage
   - Maintenance costs, including labor, lamps, ballasts
2. Proposed system — should consist of the following:
   - Number and wattage of fixture operating
   - Demand in kW
   - Burning hours
   - kWh usage
   - Maintenance costs, including labor, lamps, ballasts
3. Net results
   - kW demand reduction
   - kWh reduction
   - Electrical savings
   - Cooling load savings
   - Heating load increase
   - Maintenance cost increase or decrease
   - Utility rebate
   - Manufacturer rebate
   - Scheduled maintenance
   - Cost to complete the project

8.3.2 Evaluating methodology

Energy effectiveness in an illuminating system carries a broader significance than energy efficiency of the various components which make up the system. Energy effectiveness can only be achieved by applying and implementing the following:

1. Accurate determination of the seeing task requirements
2. Proper design approach to avoid any of the pitfalls to be introduced in the work environment
3. Optimum selection of the energy-efficient components including suitable controls for the proposed system
4. Appropriate integration of the daylighting into the system wherever applicable

Energy management and energy-effective design have a tremendous impact on cost. The final decision as to which system to install depends heavily on the cost. These costs should include not only the initial cost of the installation, but also the operating and maintenance costs. With the rise in energy cost and inflation in all sectors of the economy, an inexpensive system (low initial cost) could cost the owner many times more to operate and maintain.

8.3.3 Cost analysis

There are a number of methods of economic analysis by which the choices available may be reviewed. Types of analysis presently in use are the payback period, the internal rate of return, present value, savings investment ratio, and life-cycle costing. With the rapid increase in the cost of energy in recent years an inflationary factor is critical to an analysis of operating costs. Two of the methods, namely, payback period and life-cycle costing, will be discussed more fully in the following.

Payback period method

A major lamp manufacturer designed an energy cost management analyzer that contains all essential lamp information and fixture data, and provides parallel working spaces for the existing lighting system and the new energy-effective lighting system, side by side. Each contains items such as annual energy cost per fixture, annual cleaning cost per fixture, and so on. The total of these items is equal to the annual operating cost for the existing or the proposed system. From the difference between the two systems, one will be able to evaluate return on investment (ROI). Figure 8.1 shows such an analyzer.

There are many other forms that one can use to make a retrofitting study. However, the form prepared by IES and/or major lighting manufacturers is recommended to use for its thoroughness and completeness in essential information required for evaluation. Figure 8.2 shows such a form. Today many lamp and luminaire manufacturers offer computer services for evaluating lighting system alternatives.

Life-cycle costing analysis

Life-cycle costing (LCC) analysis is one technique which allows consideration of all the relevant economic consequences of design decisions in terms of money spent today (present cost) or money required during the life of the structure (annual cost). So LCC is the evaluation of a proposal over a reasonable time period considering all pertinent costs and the time value of the money. The evaluation can take the form of a present value analysis or uniform annual cost analysis. More sophisticated analysis would include sinking fund and rate of return on extra investment. This method of analysis is not intended to be a detailed study of various systems. It is to be used by practicing engineers as a guide for comparing the advantages of alternative design cases.

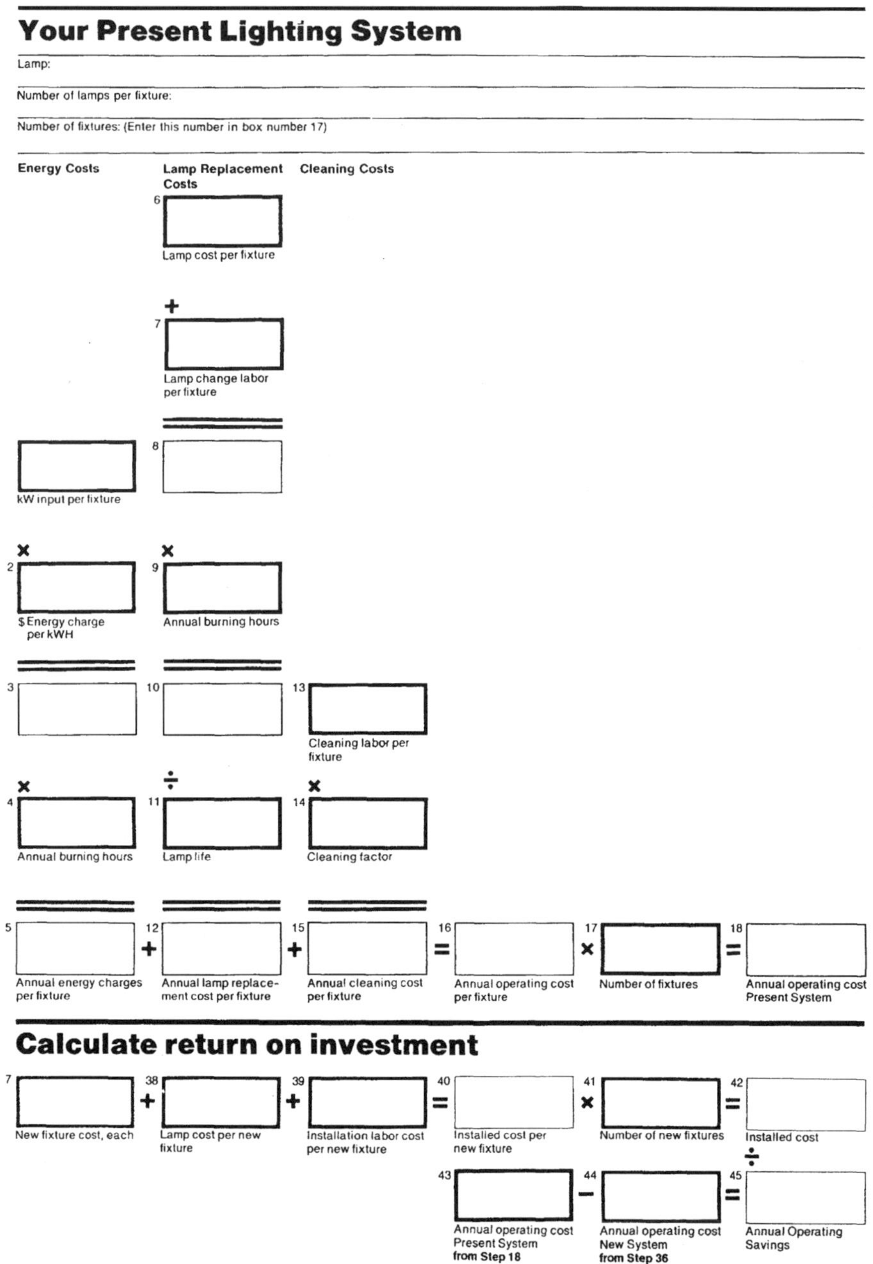

Figure 8.1(a) Lighting energy cost-management analyzer.

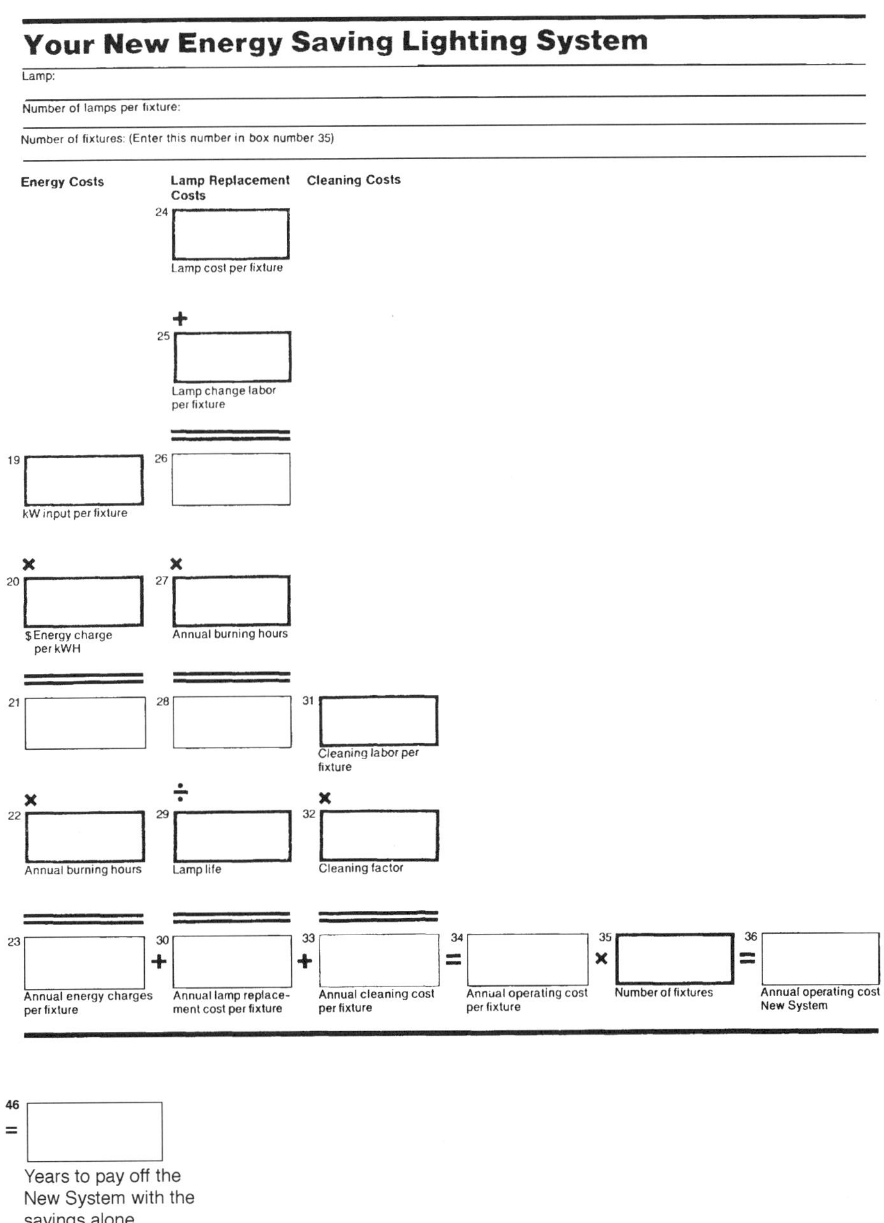

Your New Energy Saving Lighting System

Lamp:

Number of lamps per fixture:

Number of fixtures: (Enter this number in box number 35)

Energy Costs

19 kW input per fixture

× 20 $ Energy charge per kWH

= 21

× 22 Annual burning hours

= 23 Annual energy charges per fixture

Lamp Replacement Costs

24 Lamp cost per fixture

+ 25 Lamp change labor per fixture

= 26

× 27 Annual burning hours

= 28

÷ 29 Lamp life

= 30 Annual lamp replacement cost per fixture

Cleaning Costs

31 Cleaning labor per fixture

× 32 Cleaning factor

= 33 Annual cleaning cost per fixture

23 + 30 + 33 = 34 Annual operating cost per fixture × 35 Number of fixtures = 36 Annual operating cost New System

46 = Years to pay off the New System with the savings alone

Figure 8.1(b) Continued from Figure 8.1(a).

		Lighting System Parameter	Base	II
Basic Data	1.	Rated initial lamp lumens per luminaire		
	2.	Rated lamp life (hours) at ________ hours per start		
	3.	Group replacement interval (hours)		
	4.	Average watts per lamp		
	5.	Input watts per luminaire (including ballast losses)		
	6.	Coefficient of utilization		
	7.	Ballast factory (fluorescent)		
	8.	Lamp depreciation factor		
	9.	Dirt depreciation factor		
	10.	Effective maintained lumens per luminaire (1 • 6 • 7 • 8 • 9)		
	10A.	Average footcandles on work surface (10 ÷ ft^2/luminaire)		
	11.	Relative number of luminaires needed for equal maintained footcandles (10 of base system ÷ 10 of system compared)		
Initial Costs	12.	Net cost of one luminaire		
	13.	Wiring and distribution system cost per luminaire		
	14.	Installation labor cost per luminaire		
	15.	Net initial lamp cost per luminaire		
	16.	Total initial cost per luminaire (12 + 13 + 14 + 15)		
	17.	Annual owning cost per luminaire (15% of 12 + 13 + 14)		
	18.	Relative initial cost for equal maintained footcandles (16 • 11 of system compared ÷ 16 of base system)		
Operating Costs	19.	Burning hours per year		
	20.	Number of lamps group replaced per year (19 • = lamps/unit ÷ 3)		
	21.	Number of interim spot replacements (20 • = burn outs in GR interval)		
	21A.	Number of lamps spot replaced per year — No group relamping (19 • lamps/unit ÷ 2)		
	22.	Replacement lamp cost per year (20 or 21A • net lamp cost)		
	23.	Labor cost for group replacements (20 • group labor rate/lamp) at $ ________ / lamp		
	24.	Labor cost for spot replacements (21 • spot labor rate/lamp at $ ________ / lamp		
	25.	Cost of cleaning per luminaire per year		
	26	Annual energy cost per year (5 • 19 • ¢/kWH ÷ 100 000 at ________ ¢ kWH		
	27.	Total annual operating cost per luminaire (22 + 23 + 24 + 25 + 26)		
	28.	Relative annual operating cost for equal maintained footcandles (27 • 11 of system compared ÷ 27 of base system)		
Total	29.	Total annual cost — owning and operating — per luminaire (17 + 27)		
	30.	Relative total annual cost for equal maintained footcandles (29 • 11 of system compared ÷ 29 of base system)		

Figure 8.2 A typical lighting cost analysis form.

A more detailed analysis taking into account other items such as future costs could be performed. Figure 8.3 shows an outline of one method of determining costs, called "life-cycle cost analysis".

Life cycle cost analysis for ________ ft² ____________________

	Luminaire ________	Luminaire ________
	Layout ________	Layout ________
A. Lighting and air conditioning installed costs (initial)		
1. Luminaire installed costs: luminaire, lamps, material, labor	$________	$________
2. Total kW lighting:	________kW	________kW
3. Tons of air conditioning required for lighting: (3.41 × kW/12)	________tons	________ton
4. First cost of air-conditioning machinery: @ $______/ton	$________	$________
5. Reduction of first cost of heating equipment:	$________	$________
6. Other differential costs:	$________	$________
	$________	$________
	$________	$________
	$________	$________
7. Subtotal mechanical and electrical installed cost:	$________	$________
8. Initial taxes:	$________	$________
9. Total costs:	$______ (A1)	$______ (B1)
10. Installed cost per square foot:	$________	$________
11. Watts per square foot of lighting:	________watts	________watts
12. Salvage (at *y* years):	$______ (As)	$______ (Bs)
B. Annual power and maintenance costs		
1. Lamps: burning hours × kW × $/kWh	$________	$________
2. Air conditioning: operation-hours × tons × kW/ton × $/kWh	$________	$________
3. Air conditioning maintenance: tons × $/ton	$________	$________
4. Reduction in heating cost fuel used: ______	$________	$________
5. Reduced heating maintenance: MBtu × $/MBtu	$________	$________
6. Other differential costs:	$________	$________
	$________	$________
	$________	$________
	$________	$________
7. Cost of lamps: (No. of lamps ______ @ $______/lamp per *N*) (Group relamping every *N* years, typically every one, two or three years, depending on burning schedule.)	$________	$________
8. Cost of ballast replacement: (No. of ballasts ______@ $______/ballast per n) (n = number of years of ballast life.)	$________	$________
9. Luminaire washing cost: No. of luminaires ______@ $______ each. (Cost to wash one luminaire includes cost to replace *or* wash lamps.)	$________	$________
10. Annual insurance cost:	$________	$________
11. Annual property tax cost:	$________	$________
12. Total annual power and maintenance cost:	$______ (Ap)	$______ (Bp)
13. Cost per square foot:	$________	$________

Notes on analysis

A. 1. An estimate is prepared for material and labor of the installation.

2. In the example that follows a 40-W rapid-start lamp with ballasts loss is considered one 48-W load, and the 150-W HPS with ballasts is considered 175-W.

4. First cost of machinery will vary from $1000 to $2000/ton. Use the same value for both systems.

Figure 8.3 Life-cycle cost analysis form.

8.3.4 Examples of cost analysis

A comprehensive cost analysis comparing an existing mercury lighting system with four alternative systems was made for an industrial plant. It demonstrated a method for selecting a best system which is not only more energy effective, but delivers better quality illumination at the same time. Table 8.1 exhibits the comparative cost analysis. Since the study was done several years ago, the material and labor cost figures contained therein may not be up to

Table 8.1 Comparative Cost Analysis for Five Different Lighting Systems

	Existing 400-W Mercury System	250-W Metal-Halide System	400-W Metal-Halide System	250-W HPS System	400-W HPS System
Number of luminaires required	108	97	68	70	42
Luminaire spacing (square grid), ft	9.62	10.15	12.13	11.95	15.43
Initial lamp lumens per lamp	22,500	20,500	34,000	30,000	50,000
Lamp lumen depriciation factor	0.78	0.83	0.75	0.90	0.90
Estimated lamp life, h	24,000	10,000	15,000	24,000	24,000
Average lamp replacements per year	18	38.8	18.13	11.67	7
Lamp net cost, dollars per lamp	10.23	23.55	22.35	38.40	36.00
Luminaire input watts	450	285	460	300	475
Average watts per ft^2	4.9	2.8	3.1	2.1	2.0
Total connected load, kw	48.6	27.65	31.28	21	19.95
Luminaire per unit cost, dollars	0	68	85	145	150
Installation labor per unit, dollars	0	36	36	36	36
Installation cost summary					
Luminaire cost, dollars	0	6,596.00	5,780.00	10,150.00	6,300.00
Initial lamp cost, dollars	0	2,284.35	1,519.80	2,688.00	1,512.00
Installation labor cost, dollars	0	3,492.00	2,448.00	2,520.00	1,512.00
Total installation costs, dollars	0	12,372.35	9,747.80	15,358.00	9,324.00
Annual operating cost summary					
Lamp cost, dollars	184.18	913.74	405.28	448.00	252.00
Maintenance labor, dollars	180.00	288.00	181.33	116.67	70.00
Energy cost, dollars	7,776.00	4,423.20	5,004.80	3,360.00	3,192.00
Total annual operating cost, dollars	8,140.14	5,724.94	5,591.41	3,924.67	3,514.00
Relative operating cost, percent	100.00	70.33	68.69	48.21	43.17

Table 8.1 (continued) Comparative Cost Analysis for Five Different Lighting Systems

	Existing 400-W Mercury System	250-W Metal-Halide System	400-W Metal-Halide System	250-W HPS System	400-W HPS System
Total annual cost summary					
Annual owning cost, dollars	0	1,513.20	1,234.20	1,900.50	1,171.80
Owning and operating costs, dollars	8,140.14	7,238.14	6,825.61	5,825.17	4,685.80
Relative owning and operating cost, percent	100.00	88.92	83.85	71.56	57.56
Annual cost per fc/ft², dollars	0.8107	0.7216	0.6832	0.5819	0.4681
Lighting investment payback summary					
Annual operating cost, dollars	8,140.14	5,724.94	5,591.41	3,924.67	3,514.00
Operating cost savings, dollars	0	2,415.20	2,548.73	4,215.47	4,626.14
Total new investment, dollars	0	12,372.35	9,747.80	15,358.00	9,324.00
Simple investment payback interval, years	0	5.12	3.82	3.64	2.02
Simple return on investment, percent	0	19.52	26.15	27.45	49.62
Adjusted discounted investemnt payback interval, months	0	90.4	60.7	57.0	28.3
Summary of costs over next 20 years					
Net lamp costs, dollars	3,682.80	17,132.62	7,345.70	7,616.00	4,284.00
Lamp replacement labor costs (at $10 per lamp), dollars	3,600.00	7,760.00	3,626.67	2,333.33	1,400.00
Energy consumption, kWh	3,888,000	2,211,600	2,502,400	1,680,000	1,596,000
Total energy costs, dollars	155,520.00	88,464.00	100,096.00	67,200.00	63,840.00
Total initial costs, dollars	0	12,372.35	9,747.80	15,358.00	9,324.00
Total 20-year life-cycle costs, dollars	162,802.80	125,728.97	120,816.17	92,507.33	78,848.00

Note: Basis 10,000 ft² manufacturing area illuminated to 100 fc 4000 burning hours per year, effective electrical energy rate (including demand and other charges) 4 cent/kWh, average dirt conditions, 20-year amortization at interest rate of 10%.

date; however, the relative cost rankings should still hold true for evaluation purposes. Figure 8.4 shows an economic analysis for two different lighting systems in a 30 × 30-ft room to illustrate the application of the life-cycle costing method; it is not intended to show the benefits of one system over another.

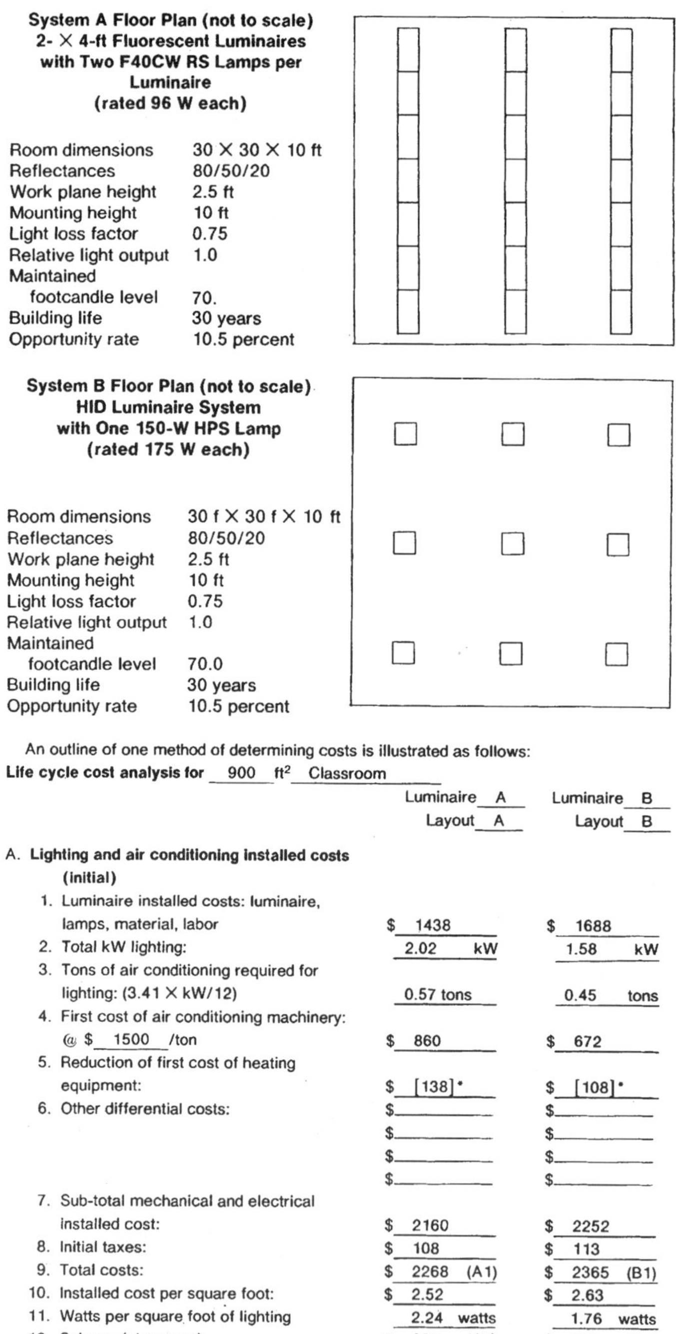

System A Floor Plan (not to scale)
2- X 4-ft Fluorescent Luminaires with Two F40CW RS Lamps per Luminaire
(rated 96 W each)

Room dimensions	30 X 30 X 10 ft
Reflectances	80/50/20
Work plane height	2.5 ft
Mounting height	10 ft
Light loss factor	0.75
Relative light output	1.0
Maintained footcandle level	70.
Building life	30 years
Opportunity rate	10.5 percent

System B Floor Plan (not to scale)
HID Luminaire System
with One 150-W HPS Lamp
(rated 175 W each)

Room dimensions	30 f X 30 f X 10 ft
Reflectances	80/50/20
Work plane height	2.5 ft
Mounting height	10 ft
Light loss factor	0.75
Relative light output	1.0
Maintained footcandle level	70.0
Building life	30 years
Opportunity rate	10.5 percent

An outline of one method of determining costs is illustrated as follows:

Life cycle cost analysis for 900 ft² Classroom

	Luminaire A Layout A	Luminaire B Layout B
A. **Lighting and air conditioning installed costs (initial)**		
1. Luminaire installed costs: luminaire, lamps, material, labor	$ 1438	$ 1688
2. Total kW lighting:	2.02 kW	1.58 kW
3. Tons of air conditioning required for lighting: (3.41 X kW/12)	0.57 tons	0.45 tons
4. First cost of air conditioning machinery: @ $ 1500 /ton	$ 860	$ 672
5. Reduction of first cost of heating equipment:	$ [138]*	$ [108]*
6. Other differential costs:	$	$
	$	$
	$	$
	$	$
7. Sub-total mechanical and electrical installed cost:	$ 2160	$ 2252
8. Initial taxes:	$ 108	$ 113
9. Total costs:	$ 2268 (A1)	$ 2365 (B1)
10. Installed cost per square foot:	$ 2.52	$ 2.63
11. Watts per square foot of lighting	2.24 watts	1.76 watts
12. Salvage (at *y* years):	$ 227 (As)	$ 237 (Bs)

Figure 8.4(a) Economic comparison analysis for two lighting systems for a classroom.

B. Annual power and maintenance costs

1. Lamps: burning hours × kW × \$/kWh	\$ 314	\$ 246
2. Air conditioning operation hours × tons × kW/Ton × \$/kWh	\$ 72	\$ 56
3. Air conditioning maintenance: tons × \$/ton	\$ 86	\$ 67
4. Reduction in heating cost fuel used: coal	\$ [23]*	\$ [18]*
5. Reduced heating maintenance: MBtu × \$/MBtu	\$ [14]*	\$ [11]*
6. Other differential costs:	\$	\$
	\$	\$
	\$	\$
	\$	\$
7. Cost of lamps: (No. of lamps $^{42}_{\underline{9}}$ @ \$ $^{1.08}_{\underline{34.00}}$ /lamp per N) $^{3\ yrs.}_{6\ yrs}$ (Group relamping every *N* years, typically every one, two or three years, depending on burning schedule.)	\$ 15	\$ 51
8. Cost of ballast replacement: (No. of ballasts $^{21}_{\underline{9}}$ @ $^{12\ yrs}_{12\ yrs}$ \$ $^{12}_{\underline{63}}$ /ballast per n) (n = number of years of ballast life.) *Labor A = 0.8 hrs, B = 1.0 hrs; Rate = 14.50/hr*	\$ 41	\$ 58
9. Luminaire washing cost: No. of luminaires $^{21}_{\underline{9}}$ @ \$ $^{7.25}_{\underline{7.25}}$ each. $^{1yr}_{1yr}$ (Cost to wash one luminaire includes cost to replace *or* wash lamps.) *Labor A & B = 0.5 hr; rate = 14.50/hr*	\$ 152	\$ 65
10. Annual insurance cost:	\$ 27	\$ 28
11. Annual property tax cost:	\$ 136	\$ 142
12. Total annual power and maintenance cost:	\$ 806 (Ap)	\$ 684 (Bp)
13. Cost per square foot:	\$ 0.90	\$ 0.76

* Bracket indicates negative values.

PAYOUT (Example)

(Sys. A) = (\$2268 − \$227) = \$2041

(Sys. B) = (\$2365 − \$237) = \$2128

$X_1 = (\$2128 - \$2041) = \$87$

$X_2 = (\$806 - \$684) = \$122$

$$X_3 = \frac{122}{(.105 \times 87)} = 13.36$$

$$a = 1.105$$

$$b = \frac{13.36}{12.36} = 1.081$$

$$y = \frac{\ln 1.081}{\ln 1.105}$$ = 0.78 years before Sys. B begins to be the more economical system. □

Figure 8.4(b) Continued from Figure 8.4(a).

References

ANSI/IEEE 739-1995, Energy Management in Industrial and Commercial Facilities.

Chen, K., The energy oriented economics of lighting systems, *IEEE Trans. Ind. Appl.*, pp. 62–68, January/February, 1976.

Chen, K. and Guerdan, E.R., Resource Benefits of Industrial Relighting Program, *IEEE Trans. Ind. Appl.*, May/June, 1979.

Chen, K. and Lally, W., Update: fluorescent lighting economics, *IEEE Trans. Ind. Appl.*, pp. 328–333, May/June, 1983.

Chen, K., Design Applications in Lighting Retrofits, IEEE-IAS Annu. Conf. Proc., October, 1992.

Wellinghoff, J., Winnning the Lighting Retrofit Game, EC&M, pp. 65–76, May, 1993.

chapter nine

Power quality, demand-side management, and harmonics

9.1 Introduction to power quality

Over the past decade, the electrical environment has undergone considerable changes. Our society has increased its dependence on electronic systems, which now proliferate. The dependence on electronic systems began in the 1980s with the personal computer (PC) revolution. Computing power moved out of the computer room's managed environment to just about everywhere — factory floor, desktop, etc.

In the 1990s came the network revolution with ever-increasing equipment capability. Now computing power was connected together to form networks. Networked facilities increased the exposure and potential for interference that could be coupled into the sensitive electronic equipment on which the process or activity depended.

During the early 1990s the electrical environment went through some dramatic changes:

1. PCs with their phase-neutral connected rectified power supplies changed the load characteristics of office buildings. It caused overloaded circuits, harmonic currents, and overheated electrical distribution equipment.
2. Equipment sensitivities have changed because of new levels of circuit integration. New generations of equipment were found to be sensitive to existing environment noise to which older equipment was immune.
3. Systems were assembled with parts from various sources and not designed as total systems. EMI/EMC issues were largely ignored by companies eager to capitalize on the demand for equipment and systems.
4. Large loads within buildings were changed to modern energy-efficient variable-speed drives, frequency modulators, and large UPS equipment. These devices had the potential to introduce additional interference, including harmonics, into the electrical environment.

5. Energy-efficient lighting systems were installed at a wholesale rate. The focus was on saving energy and little care was given to experts in the marketplace with the organization and resources to manage the end users' electrical energy needs.

9.2 Demand-Side Management and the Energy Policy Act

Demand-side management (DSM) and other energy conservation programs are important and offer a growing market. Although they appear beneficial when standing alone, they can have a negative impact on the quality of power and the overall reliability of the end users' electrical system.

The industrial users are of greater concern. The problem that both DSM proponents and electric utilities share is that they view the cost of a delivered kW to be of singular overriding significance. They forget that although a reduction in electrical power consumption may reduce that cost of all delivered goods, a reduction in yield due to power quality problems will significantly increase the cost of the same goods.

The general observation is that electric utilities are not doing a good job of identifying, segmenting, or positioning products in the coming era of deregulation. The majority of discussions will be focused on how the largest industrial customers are going to be captured by the lowest cost power suppliers. To combat this intrusion into the end users' power supply business, utilities must offer unregulated products and services that address the unique needs of each customer.

Once the free market distinguishes a kWh on the basis of cost, quality and security, etc. and appreciates the distinction in price, consumers will conserve or consume based on their ability to pay just like they do with any other commodity.

Does DSM have a place in such a market? Most of the evidence to date confirms that savings are either less than expected and/or that savings are exceedingly difficult to measure. Of course there is no good reason to suppress electricity consumption if overall benefits are positive.

DSM programs are relatively recent. Many of the implications are just being realized and played out. One is to include all the costs and impacts. For example, the costs to recycle fluorescent lamps, because they include compounds like mercury and other metals, are now being factored into some programs.

Implications of the Energy Policy Act of 1992 for DSM efforts

The Energy Policy Act of 1992 will affect utilities, regulators, and DSM professionals in a number of ways:

1. The DSM community can play a critical role in implementation. The legislation only goes so far in mandating efficiency improvements. In

the case of financial incentives for utilities, the legislation only requires consideration of these goals and offers co-funding for regulatory proceedings. Utilities and energy efficiency advocates can support strong building codes and equipment efficiency standards as these portions of the Act are implemented. Utilities should consider taking a more active position given the energy savings potential and the fact that the savings are "free" to utilities.

2. Funding will be required to implement the Act. Utilities, states, and regulators could help to advocate full funding for the efficiency provisions. Full funding implies a significant increase in DOE's energy efficiency budget. Such increases could be offset if needed by a small reduction in DOE's nuclear and fossil fuel programs.
3. Utility DSM programs should be adjusted as the Act is implemented. For example, certain energy-efficient products such as energy-saving lamps, energy-efficient motors, and low-flow faucets and showerheads will become the norm once the new equipment standards take effect. Utilities should then remove these products from their incentive programs, or at least ensure that the products they are promoting are significantly more efficient than the minimum standards. Likewise, many states will revise their commercial building code if the Act works as intended. If so, utilities should revise their DSM programs for new construction so that they are promoting building practices that significantly exceed minimum code.
4. Utilities and regulators should take the energy efficiency provisions into account as they forecast electricity and natural gas demand and plan new facilities. The Act should enable utilities to avoid or defer many new power plants. In fact, society will not realize the full economic benefits of the legislation unless these plants are avoided.

9.3 *Power disturbances defined*

The IEEE Standard 1159-1995, "Recommended Practice for Monitoring Electrical Power Quality," defines various disturbances as interruptions, sags and swells, long-duration variations, impulsive transients, oscillatory transients, harmonic distortion, voltage fluctuations, and noise.

Interruptions

An interruption is the complete loss of supply voltage or load current. Within this definition there are three types of interruptions that are characterized by their duration: momentary, temporary, and long term. The momentary interruption is a complete loss of supply voltage or load current having a duration between 0.5 cycle and 3 s. The temporary interruption is the complete loss lasting between 3 s and 1 min. The long-term interruption or outage has a duration of more than 1 min.

Normally interruptions result in the operation of a system-protective device, such as a fuse or automatic breaker, that isolates the source of the system default.

Sags and swells

They are grouped by IEEE 1159 under the general heading of Short-Duration Variations, which are disturbances with a duration of less than 30 cycles. A sag is defined as a short-duration, temporary voltage drop that lasts between 0.5 and 30 cycles and has a typical magnitude in the 0.1 to 0.9 per unit range. A swell is similarly defined as a short-duration, temporary voltage rise that lasts between 1.1 and 1.8 per unit.

Common causes of sags include the starting of large loads and remote fault clearing. For swells, high impedance neutral connections, sudden load reductions, and a single-phase fault on a three-phase system are common sources. The result can be data errors, flickering of lights, degradation of electrical contacts, semiconductor damage in electronics, and insulation degradation.

Long-duration variations

It can be either overvoltages or undervoltages. The key characteristic for this type of disturbance is that it lasts longer than 1 min. Many of the same results are experienced with that of sags and swells. Dimming of lights, data errors, electronics damage or reduction of life, and insulation degradation can occur. Poor voltage regulation often is the source of long-duration variations.

Impulsive transients

Much of the misunderstanding between utility companies and industry personnel covers impulsive transient disturbances. Many different terms have been used to describe impulsive transients, such as bump, glitch, power surge, and spike. The IEEE defines an impulsive transient as a sudden, nonpower frequency change in the steady-state condition of the voltage, current, or both. A key point in recognizing this type of disturbance is its undirectional polarity. Sources of impulsive transients include lightning, poor grounding, the switching of inductive loads, the normal operation of electronic loads, and fault clearing. The results from such a transient can range from the loss or corruption of data to physical damage of equipment.

Oscillatory transients

They are a common cause of equipment shutdown. The most common source of this type of disturbance is capacitor switching. An oscillatory is defined as a voltage or current whose instantaneous value changes polarity rapidly. These transients can be categorized by their frequency, with components less than 5 kHz; medium frequency, with components between 5 and 500 kHz;

or high frequency, with components between 500 kHz and 5 Mhz. The most recognized problem associated with capacitor switching is the nuisance tripping of adjustable speed drives (ASD).

Harmonic distortion

It is observed when sinusoidal voltages or currents have frequencies that are integer multiples of the fundamental frequency being supplied. This distortion is continuous, and the most common result is unwanted heating in the electrical system. Examples include transformer overheating and neutral conductor overheating.

It is interesting to note that some of the equipment that are most sensitive to power quality disturbances are often the same equipment that generate harmonics. Equipment such as ASDs, computer power supplies, UPS equipment, and other power electronics create harmonic currents. Harmonic currents generate harmonic voltages as they pass through the system impedance. In addition to power electronics, arcing equipment, such as arc furnaces and welders, are major contributors in the harmonic arena.

Voltage fluctuations and noise

Voltage fluctuations are defined as a series of random voltage changes, the magnitude of which does not exceed the voltage range of 0.95 to 1.05 per unit. These changes are the result of rapidly varying load, such as the use of an arc furnace. Voltage fluctuations and flicker are terms often used to denote the same type of disturbance; however, it should be noted that flicker is actually the visible result of voltage fluctuations as seen in lighting. The two terms should not be used interchangeably.

Noise is unwanted electrical signals that are superimposed upon the power system voltage or current. This disturbance is continuous and exhibits broadband frequency components.

Figure 9.1 shows wave representation of some of the above-discussed power disturbances.

In conclusion, the widespread use of electronics has raised the awareness of power quality and its effect on end-use equipment. Economical solutions are available to mitigate the effects of power quality disturbances.

The following sections will be devoted to the harmonics problems, which are considered to be the most serious among many other disturbances as discussed above.

9.4 Harmonics and its effect on the equipment performance

9.4.1 Harmonic currents

Harmonic currents do their dirty work within the facility where they are generated. The effects of harmonic currents concern utilities to such an extent

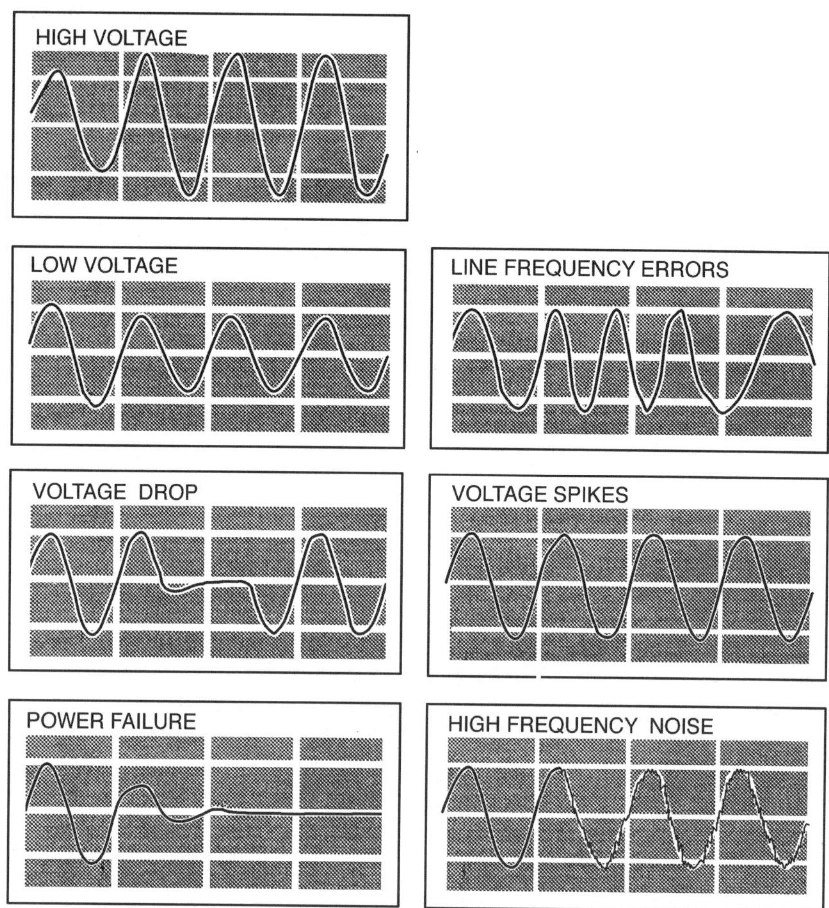

Figure 9.1 Wave representation for power disturbances.

that they must account for currents circulating in delta-wye distribution transformers. In wye-connected three-phase power systems, the presence of harmonic components can cause excessive current flow in the neutral line. The neutral current can be large enough to damage insulation and create conditions favorable to combustion.

Harmonic currents distort the line voltage waveform while bothering a number of power users only from an aesthetic point of view. Such distortion can result in degradation of effective harmonic fluxes within motor armature, hot spots within transformer windings, and improper operation of electronic controls and clocks.

9.4.2 Types of energy-efficient equipment which generates harmonics

All nonlinear loads in a commercial and industrial facility can be grouped into three general categories:

1. Electronic power supplies — This category includes all modern equipment used in the office environment: personal computers, printers, typewriters, copiers, etc. All contain power supplies with similar characteristics. They are typically single phase and require 120 V power supply.
2. Fluorescent lighting — Conventional magnetic ballasts are rapidly being replaced by electronic ballasts. Harmonic generation characteristics of electronic ballast vary over a considerable range. Fluorescent lighting can be supplied at either 120 or 480 V (277 V line to neutral) system.
3. Adjustable speed drive for HVAC — Induction motors which were traditionally used for chillers and fan drives are being replaced with adjustable speed motor drives (ASD). The variable nature of HVAC loads makes them well suited for ASDs because of large potential energy savings. These loads are typically three phase and may be connected directly to the 480-V supply, or through isolation transformers.

Although ASDs, uninterruptible power supplies, PCs, and fluorescent lighting systems all contribute to the generation of harmonics, utilities appear to consider the electronic ballasts as the major problem. Manufacturers developed electronic ballast with harmonics less than 20% using the passive filtering method. By using active-type circuits, electronic ballasts are available with less than 10% total harmonic distortion (THD). Utilities are concerned with the effects of harmonics on system and equipment performance. Electronic components of any ballast may cause a phase shift between the current and voltage which cannot be corrected in the line, but must be addressed in the circuit design of the ballast. The low power factor for electronic ballast is caused primarily by current distortion. In Figure 9.2 the distorted curve is an example of the wave shape of the current drawn by a ballast with THD of about 33%.

The power factor can be corrected to well above 0.90 for electronic ballast by improving the wave shape. This can be accomplished with passive circuits or expensive active circuits. Distorted wave shapes contain components with frequencies that are multiples of the fundamental frequency. These higher frequency components are known as harmonics. Fluorescent systems, when operated with electronic ballasts, typically have THD from 5 to 30%.

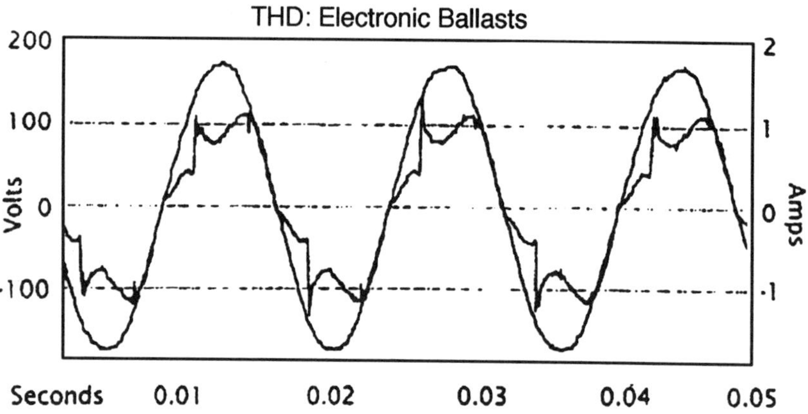

Figure 9.2 Smooth voltage wave vs. distorted current wave by harmonics. The smoother curve is a voltage wave with near-zero distortion, typically supplied to ballasts. The distorted curve is an example of the wave shape of the current drawn by a ballast with a total harmonic distortion of about 33%.

Compact fluorescent lighting may affect power quality in two aspects:

1. Power factor (PF) — Independent testing of compact fluorescent lamps found the normal PF ranges between 0.41 and 0.61 in base-up and base-down orientation. Some CFL products had PF exceeding 0.90 in a base-up position. With normal PF CFLs are used to replace incandescent lamps of comparable light output. The reduced PF does not cause a current overload in the existing system because the reduced active power more than compensates for the reduced PF. Figure 9.3 shows typical current wave shapes of two normal PF CFLs and an incandescent lamp. Two aspects of the current wave shape reduce PF: phase displacement and harmonic distortion.
2. Electromagnetic interference (EMI) — Electronic devices often employ power supplies that can generate EMI. Electronically ballasted CFL must comply with FCC regulations requiring the amount of conducted EMI that they may produce. Radiated EMI usually occurs in two frequency bands. The first is between 10 and 100 kHz which is below the AM radio band. This problem is not commonly observed. The second frequency band includes infrared radiation. Modulated infrared radiation from CFL is anecdotally reported to interfere with the operation of remote controllers of audio-voiced equipment, such as television and video cassette recorders.

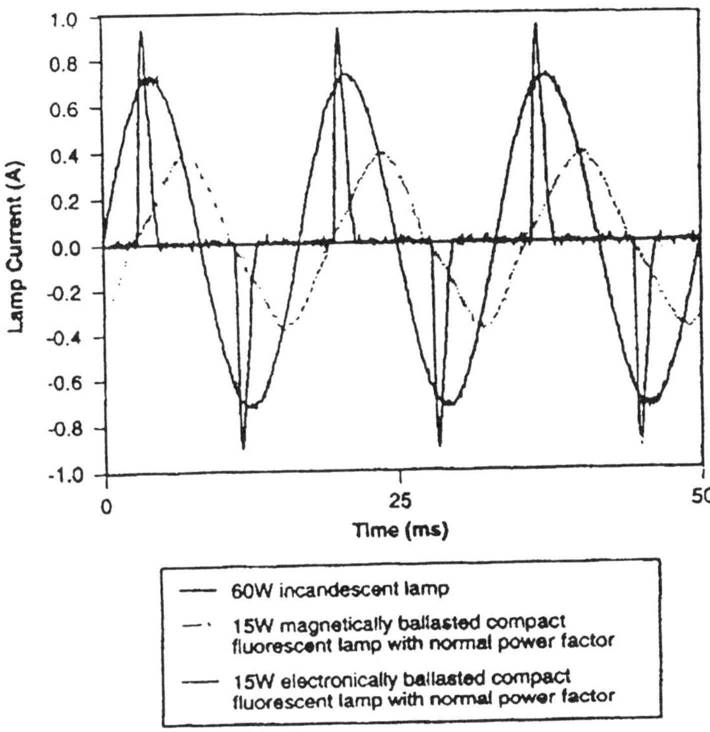

Figure 9.3 Power quality comparison of incandescent vs. compact fluorescent lamps with different ballasts. An incandescent lamp draws a current that has the same shape as, and is in phase with, the supplied voltage. A magnetically ballasted compact fluorescent lamp draws a current that lags the voltage and is slightly distorted. An electronically ballasted compact fluorescent lamp draws a current that is highly distorted and slightly leads the voltage.

9.4.3 Terminal characteristics of diode bridge rectifiers

The power converters are all connected to the AC power system through a single- or three-phase diode bridge rectifier. This is considered as the culprit of all evils for causing power quality problems.

Terminal characteristics of diode bridge rectifiers can be summarized as follows:

1. Single-phase electronic power supplies are becoming a dominant load in commercial buildings. Almost all electronic power supplies contain a diode bridge rectifier, which supplies DC power to large DC filter

capacitors. The rectified filter DC potential is then electronically regulated by a switch-mode DC/DC converter before it is passed to microelectronic circuitry. The combination of diode-bridge rectifier and DC filter capacitor results in a current waveform with high peak values. The peaks are centered around the peak of the AC terminal voltage. Input current from the AC power line contains a high amount of odd harmonic components, with the largest components occurring at the 3rd, 5th, and 7th harmonic frequencies. A group of electronic power supplies fed from the same branch distribution circuit will appear much like a single large load. An observable effect in the input current to an aggregate power supply load is some reduction in harmonic content at higher frequencies, with nearly arithmetic addition of harmonics at lower frequencies. The effect on the input current waveform is a broadening of the "pulse" width.

2. Electronic ballasts also contain diode bridge rectifier circuits which supply a DC voltage to an inverter, which feeds the fluorescent lamp. DC filter capacitors are smaller than those found in electronic power supplies. This in combination with series inductance on the AC side of the rectifier for limiting conducted EMI generation reduces input current distortion. Electronic ballasts generally exhibit a wide range of harmonic current characteristics, from a low of 10% THD to a high of about 50% THD. Harmonics generated by a single type of electronic ballast add almost arithmetically in branch distribution circuit.
3. ASD for HVAC applications — Small ASDs used in HVAC applications also contain a diode bridge rectifier and DC filter capacitor. The DC voltage is supplied to a pulse-width modulated (PWM) inverter, which in turn provides variable frequency and variable voltage to an AC induction motor. Operational characteristics of the input rectifier are similar to those found in electronic power supplies or electronic ballasts, but the three-phase connection results in two current "pulses" during each half-cycle of AC voltage. The effect on the harmonic generation under balanced conditions is the elimination of the third harmonic component.

9.4.4 Fundamentals of harmonic distortion and applications

The current waveform from the line into the system is 60 Hz periodic and therefore has a Fourier spectrum:

$$i(t) = \sum_{n=1}^{inf} a_n \sin(nwt) + \sum_{n=1}^{inf} b_n \cos(nwt)$$

where

$$a = 1/\pi \int_0^{2\pi} i(t) \sin(nwt)\, d(wt)$$

$$b = 1/\pi\int_0^{2\pi} i(t)\cos(nwt)\, d(wt)$$

The magnitude (rms) of the nth harmonic can be expressed by

$$I(rms) = (1/2)\sqrt{a_n^2 + b_n^2}$$

$$I_{total}(rms) = \sqrt{I_{1(rms)}^2 + I_{2(rms)}^2 + \ldots + I_{n(rms)}^2}$$

The THD is the ratio in percent of the rms value of the harmonics compared to the rms value of the fundamental.

$$THD(\%) = \frac{100 \times \sqrt{I_{2(rms)}^2 + \ldots I_{n(rms)}^2}}{I_{1(rms)}}$$

Harmonic voltage distortion

Voltage produced by utility generators are remarkably free of harmonic distortion. At the same time, ambient levels of harmonic voltage distortion are commonly detectable at utilization levels. The cause of those distortions at utilization and distribution system levels can be attributed to distorted current drawn by nonlinear customer loads.

$$V_h = I_h Z_h$$

Total harmonic distortion of the voltage can be determined from superposition of the results of computations at each frequency.

The impedance of the power system at harmonic frequencies can be calculated as:

$$X_h = h\frac{kV^2}{kVA}Z_T(pu)$$

$$X_h = hZ_T\ (\%)\ (\text{in \% on transformer base})$$

Where the zero-sequence triplen harmonics are trapped in the delta winding, positive sequence harmonic current will experience a phase shift of –30° times the harmonic number from secondary to primary. Negative sequence harmonics are shifted in the opposite direction, +30° times the harmonic order.

Phase shift created by delta-wye transformers can sometimes reduce the total harmonic current at the service entrance.

Harmonic currents in neutral conductors

Another well-known consequence of the characteristics of nonlinear loads is harmonic current loading of three-phase 4-wire circuit neutral conductor.

Neutral current loading in three-phase circuits with linear loads is simply a function of the load balance among the three phases.

In circuits with nonlinear load, zero sequence harmonic currents will add arithmetically in the neutral conductor of three-phase circuits. The third harmonic is usually the largest harmonic component in single-phase electronic power supplies or electronic ballasts.

For the distorted currents the rms value can be calculated from the harmonic components as:

$$I\ rms = \sqrt{I_1^2 + I_3^2 + I_5^2 + I_7^2 + \ldots}$$

Triplen harmonic currents are not a problem in three-phase nonlinear loads, such as ASDs, since the topology of the diode bridge rectifier circuit in effective line-to-line connection at the input prevents these currents from flowing into the power system.

For the two classes of single-phase nonlinear loads considered here, typical values of neutral current loading as a percentage of balanced phase current loading can be estimated from the harmonic spectrums provided:

$$\begin{aligned} I_{l\text{-rms}} &= \sqrt{I_1^2 + I_3^2 + I_5^2 + I_7^2 + \ldots} \\ &= \sqrt{I_1^2 + I_3^2} \\ I_{n\text{-rms}} &= \sqrt{(3I_3)^2 + (3I_9)^2 + (3I_{15})^2 + \ldots} \\ &= 3I_3^3 \end{aligned}$$

For electronic power supplies, where $I_3 = 0.7\ I_1$,

$$\frac{I_{n\text{-rms}}}{I_{1\text{-rms}}} = \frac{3(0.7)}{\sqrt{1^2 + 0.7^2}} = 1.7$$

For electronic ballasts, with $I_3 = 0.3\ I_1$,

$$\frac{I_{n\text{-rms}}}{I_{1\text{-rms}}} = \frac{3(0.3)}{\sqrt{1^2 + 0.3^2}} = 0.8$$

Effects of nonlinear load penetration

In evaluating harmonic impacts at the service entrance for nonlinear loads in industrial and commercial facilities, a primary consideration should be the percentage of the total facility load that is nonlinear. At 208/120-V panels served by 480-V transformers, the nonlinear load served by each of these transformers must be considered for evaluating both harmonic voltage distortion and transformer derating. Small amounts of nonlinear load relative to transformer capacity will not cause harmonic problems.

Harmonic current generation from nonlinear loads in industrial and commercial facilities continues to grow as factories and offices become more automated and energy-efficient equipment with nonlinear characteristics replaces less efficient linear loads. Effects of high relative levels of harmonic current generation in these facilities include:

- Harmonic voltage distortion
- Transformer and motor overheating
- Neutral conductor overloading

To sum up, all foregoing discussions on harmonics and its effects om power quality can be concisely tabulated in Table 9.1.

9.5 Transformer derating and "K" factor

True power factor

In more general form, power factor is defined as:

$$\text{P.F.} = I_1 / I_{rms} \cos(\theta)$$

This quantity is sometimes called true power factor, to distinguish it from the utility definition, known as displacement power factor.

The load discussed here exhibits poor uncorrected true power factor. However, displacement power factors for these loads are usually above 0.95, and in most cases closer to 1.0. As to the traditional corrective measure for poor "power factor", the addition of shunt capacitors will not improve the true power factor of this type of load.

For diode bridge rectifier loads, true power factor can be expressed as:

$$\text{P.F.} = \sqrt{\frac{1}{1 + \text{THD}^2}}$$

True power factor of electronic loads, where current THD may be on the order of 70 to 90%, is 0.6 to 0.75. Electronic ballasts with the current THD below 35% have true power factor greater than 0.94.

Transformer derating

Power transformers not specially designed for nonsinusoidal load currents must be derated to account for the additional winding eddy current losses from harmonic currents.

ANSI/IEEE C57.110 shows a procedure for establishing transformer capability for supplying nonsinusoidal load currents. UL is able to list

Table 9.1 Effects of Harmonics on Components of a Power Distribution System

EACH COMPONENT of a power distribution system manifests the effects of harmonics a little differently.

Component	Effects of harmonics	Reactions
Neutral conductors	In a 3-phase, 4-wire system, neutral conductors can be severely affected by nonlinear loads connected to 120-V branch circuits; if the loads are single phase, certain harmonics called triplens (odd multiples of the third harmonic: 3rd, 9th, 15th, etc.) do not cancel, but instead add together in the neutral conductor	If the system has many of these loads, the neutral current can exceed any of the phase currents — a real danger because no circuit breaker protects the neutral
Circuit breakers	*Thermal magnetic circuit breakers* use a bimetallic trip mechanism that responds to the actual heating value of the current waveform	They trip when they get too hot, protecting against harmonic current overloads
	Peak-sensing circuit breakers respond to the peak of the current waveform and do not always respond properly to harmonic currents	Because the peak of the harmonic current is usually higher than normal, the breaker may trip prematurely at a low current; if the peak is lower than normal, the breaker may fail to trip when it should
Bus bars and connecting lugs	Neutral bus bars and connecting lugs are sized to carry the full value of the rated phase current	They can become overloaded when the neutral conductors are overloaded with the additional sum of triplen harmonics
Electrical panels	Panels designed to carry 60-Hz currents can become mechanically resonant to the magnetic fields generated by higher-frequency harmonic currents	The panel vibrates and emits a buzzing sound at the harmonic frequencies
Telecommunication systems	Telecommunications systems often give the first clue to a harmonics problem; telecommunications cable is commonly run next to neutral conductors	However, triplens in the neutral conductor cause inductive interference, which can be heard on a phone line
Transformers	Some commercial buildings have a 208/120 V transformer in a delta-wye configuration; these transformers commonly feed receptacles in a commercial building; single-phase nonlinear loads connected to the receptacles produce triplen harmonics, which algebraically add up in the neutral	When neutral current reaches the transformer, it is reflected into the delta primary winding, where it circulates and causes overheating and transformer failures

Table 9.1 (continued) Effects of Harmonics on Components of a Power Distribution System

Component	Effects of harmonics	Reactions
Transformers (continued)	Or a transformer problem can occur from core loss and copper loss; transformers are normally rated for a 60-Hz phase current load only; higher-frequency harmonic currents cause increased core loss due to eddy currents and hysteresis, resulting in more heating than would occur at the same 60-Hz current	Possible burnout; these heating effects demand that transformers be derated for harmonic loads or replaced with specially designed "k-rated" transformers
Generators	Standby generators are subject to the same kind of overheating problems as transformers; because they provide emergency backup for harmonic-producing loads, they are often even more vulnerable	Besides overheating, certain harmonics produce distortion at the zero crossing of the current they waveform, which causes interference and instability for the generator's control circuits

specially constructed transformers for powering harmonic-generating loads. These transformers have a notation on their nameplates, "Suitable for non-sinusoidal current load with a K Factor not to exceed … ." Why is K factor necessary? We must examine the effects of nonlinear load on transformers:

1. Skin effect — Harmonic current can cause overheating of conductors and its insulating materials as a result of the skin effect. As the frequency increases current will tend to flow toward the conductor's surface and away from its cross section. Consequently, conductor resistance and load increase substantially. The resultant third harmonic contents in the neutral conductor will increase.
2. Eddy-current loss — When alternating current is passed through the transformer windings, an additional loss called stray loss exists. Only that portion in the transformer windings is termed "eddy-current loss". The portion outside the windings is termed "other stray losses". The total winding loss of a transformer is equal to I^2R loss plus eddy-current loss plus other stray loss. Because eddy-current loss is proportional to the square of the load current and the square of the frequency, nonsinusoidal load currents cause a very high level of winding loss which can lead to abnormal winding temperature rise.
3. Core flux density increase — Transformers are usually the largest ferromagnetic loads connected to any distribution system. Even partial saturation of a transformer's iron core will cause harmonic magnetizing current that differs greatly from a sinusoidal wave.

Calculating K factor

IEEE 57.110-1998 provides a method of calculating the additional heating, or watt loss, that will occur in a transformer when supplying a load that generates a specific level of harmonic distortion.

$$I_h\,(pu) = \%(HD)\sqrt{(THD)^2 + (100)^2}$$

where h = harmonic frequency order number (1, 3, 5, etc.)
$I_h\,(pu)$ = per unit rms current at harmonic order h
$\%\,HD\,(h)$ = percent harmonic distortion at harmonic order h
THD = total harmonic distortion

$I_h\,(pu)^2h^2$ is calculated for each of the harmonic orders and then added together. The summation called the K factor is then multiplied by the eddy-current losses, which yields the amount of increased transformer heating. (Eddy-current losses can be found by subtracting the I^2R losses from measured impedance losses.)

The K-rated transformers are rated at 600 V, ventilated dry types available in K ratings of 4, 13, 20, 30, or 50. Temperature rises of 150°, 115°, and 80°C at rated voltage are available.

Cautions for using K factors

Misunderstandings about transformer K ratings come from failure to recognize the difference between individual branch load K factor and the total load harmonics that appear at the feeder transformer terminals. Specifying an excessively high K factor (more than 13) can create potential hazard from abnormally low impedance, a typical symptom of oversized transformers. Where K-rated transformers are necessary, they should be specified as close as possible to the actual load K factor needed, and designed within the normal impedance range to provide the beneficial role of "softening" harmonics and reducing neutral current that transformers have been providing all along. It is also desirable to size transformers close to the load kVA requirements in order to make full use of the reactance in lowering harmonics and neutral currents.

9.6 Standards

Recent advancements in energy-efficient devices have prompted the utilities and customers alike to desire some standards on PF, THD, EMI, etc. limiting value to reduce the impact of devices on power quality. This important task should be the common goal for the power engineers, the illuminating engineers, and the electronic circuit design engineers to work together and develop a useful and applicable standard to safeguard the power quality of the power systems.

Table 9.2 lists the current or existing IEEE Standards on power quality and the corresponding IEC's EMC Standards covering various PQ attributes.

Table 9.2 IEEE vs. IEC Standards Concerning Power Quality

Electrical condition	Power quality concern	IEEE PQ standards	IEC EMC standards
Harmonics, voltages, and currents (resonance, telephone influence factor, notching)	Harmonic environment		IEC 1000-2-1/2
	Compatibility limits	IEEE 519	IEC 1000-3-2/4 (555)
	Harmonic measurement		IEC 1000-4-7/13
	Harmonic practices	IEEE P519A	IEC 1000-5-5
	Power component heating	IEEE/ANSI C57.110	
Undervoltages (low-voltage sags, dips)	Sag environment	IEEE 1250	IEC 38, 1000-2-4
	Compatibility limits	IEEE 1346	IEC 1000-3-3/5 (555)
	Sag measurement		IEC 1000-4-1/11
	Sag mitigation practices	IEEE 446,1100,1159	IEC 1000-5-X
	Fuse blowing/ upsets	IEEE 242-protection	IEC 364
Overvoltages (temporary overvoltages, surges, transients)	Surge environment	IEEE/ANSI C62.41	IEC-1000-2-5
	Compatibility limits		IEC-1000-3-X
	Surge measurement	IEEE/ANSI C62.45	IEC-1000-4-1/2/4/5/12
	Surge protection practices	C62 series, 1100	IEC 1000-5-X
	Insulation breakdown		IEC 664

Note: This table compares the standards of the IEEE/Americcan National Standards Institute (ANSI) and the IEC/European CENELEC (European Community Standards Organization) according to the power quality concerns they address. Standards denoted by a P for IEEE and an X for IEC are not yet published, but the need for them has been recognized. Blanks indicate that no standards exist and none are in process.

References

Chen, K., The Impact of Energy Efficient Equipment on System Power Quality, IEEE Distinguished Lectures Series #3, October, 1996.

DSM Q., Official Publication of the Demand-Side Management Society/AEE, Winter, 1993.

Ellis, R. G., Harmonic analysis of industrial power systems, *IEEE Trans. Ind. Appl.*, 32 (2), pp. 417–421, March/April, 1996.

Frank, J., The How and Why of K-Factor Transformers, EC&M, May, 1993.

Stebbins, W. and Dougherty, J., IEEE standard helps define power quality problems, *Energy User News*, October, 1997.

chapter ten

Lighting energy standards

10.1 Chronological development of the energy standards

Early in 1974, the document "Energy Conservation Guidelines for Existing Office Buildings" was published by the General Service Administration and the Public Building Services (GSA/PBS). At approximately the same time, the National Bureau of Standards (NBS) prepared the document "Design and Evaluation Criteria for Energy Conservation in New Buildings" for the National Conference of States on Building Codes and Standards (NCSBCS) covering lighting for new construction. NCSBCS asked the American Society of Heating, Refrigeration, and Air Conditioning Engineers (ASHRAE) to prepare a standard on energy conservation for new buildings based on this NBS document. ASHRAE turned to the Illuminating Engineering Society (IES), and the IES Task Committee on Energy Budgeting Procedures was formed. The work of this committee culminated in recommendations that were published by ASHRAE as Chapter 9 of the ASHRAE 90-75 and by IES as "IES Recommended Lighting Power Budget Determination Procedures EMS-1". NCSBCS also adopted it as part of its Model Building Code. Both documents established a procedure for determining a "lighting power budget".

In 1975, Congress passed Public Law 94-163, entitled "Energy Policy and Conservation Act of 1975", which was amended by Public Law 94-385, "Energy Conservation Standards for New Building Act of 1976". Public Law 94-164 makes mandatory certain lighting efficiency standards set forth in Public Laws 94-163 and 94-385.

In 1976, the Energy Research and Development Association (ERDA) contracted with NCSBCS to codify ASHRAE 90-75. The resulting document was called "The Model Code for Energy Conservation in New Building", or, more simply, the "Model Code". The Model Code has been adopted by a number of states to satisfy the requirements of Public Laws 94-163 and 94-385. In June 1976, the IES Board of Directors adopted important revisions to Chapter 9 of ASHRAE 90-75 that tightened the lighting power budget procedure to assure energy conservation. These revisions were included in 90-75R but did not become part of the NCSBCS Model Code.

ASHRAE 90-75 was cosponsored by IES and ASHRAE and submitted to the American National Standards Institute (ANSI) in late 1977 for adoption as an ANSI standard. The document that resulted was known as ANSI/ASHRAE/IES Standard 90, "Energy Conservation in New Buildings".

There have been several revisions on the ANSI/ASHRAE/IES 90-75R since then. All were included in the lighting portion of ANSI/ASHRAE/IES 90A-1980, "Energy Conservation in New Building Design", and in EMS-1-1981, "IES Recommended Lighting Power Budget Determination Procedure".

ANSI/ASHRAE/IES Standard 90-75R was by far the most popular energy conservation design standard. It has been adopted as energy codes by most states within the U.S. and serves as a model standard in many other countries. Although this standard underwent a revision in 1980, major revisions were not made to incorporate new practices and technology.

ASHRAE/IES 90.1-1989, "Energy Efficient Design of New Buildings except New Low-Rise Residential Buildings", is the third-generation document on building energy efficiency since the first publication in 1975. It sets forth design requirements for the efficient use of energy in new buildings intended for human occupancy. The requirements apply to the building envelope, distribution of energy, system and equipment for auxiliaries, heating, ventilating, air-conditioning, service water heating, lighting, and energy management. This standard is intended to be a voluntary standard which can be adopted by building officials for state and local codes. Three parallel and alternative paths for compliance are provided: prescriptive, system performance, and building energy cost budget methods. Each of these methods may be used selectively and interchangeably when determining building subsystems compliance. The interaction between the three compliance methods fulfills needs that arise during the various phases of the building process.

ASHRAE/IES 90.1 is currently being revised and updated and, when finalized, will supersede the 1989 standard. Basic requirements of the revised lighting section include lighting power limits for common interior and exterior applications, minimum requirements for lighting controls, and minimum efficacy requirements for exit signs. The revised lighting section has the following significant improvements over the 1989 version:

1. A shorter, more concise body of text that is written in code language to eliminate ambiguity and inconsistent interpretation. The overall standard has been simplified while adding a degree of flexibility to accommodate special design considerations.
2. A simplified, but more aggressive set of requirements for lighting controls specifically in the area of automatic shutoff.
3. A more stringent set of LPD values developed by industry consensus, that correlate closely to real spaces and actual lighting installations.
4. The revised standard will address major alterations and additions to existing buildings which is a large and rapidly growing segment of building activity.

10.2 Development of the lighting energy standards

In 1975, work was begun by IES on a series of six documents that deal with energy standards for existing buildings. The documents cover low-rise residential, high-rise residential, commercial, industrial, institutional, and public assembly occupancies. Several of these documents have since become ANSI standards. As a means for simplifying and shortening the EMS-1 procedure, the IES developed a Unit Power Density (UPD) procedure and published it as EMS-6 in 1980. The present LEM-1-1982, "Lighting Power Limit Determination", is a further refinement which combines EMS-6 and EMS-1 and, as such, supersedes both. LEM-1 is concerned only with the determination of lighting power limit; lighting control guidelines are contained in LEM-3, "Design Considerations for Effective Building Lighting Energy Utilization".

Since 1982, IES has published the LEM series. In addition to LEM-1 and LEM-3 are LEM-2, "Lighting Energy Limit Determination", LEM-4, "Energy Analysis of Building Lighting Design and Installation", and LEM-6, "IES Guidelines for Unit Power Density (UPD) for New Roadway Lighting Installation".

Since 1995, the IES has withdrawn LEM-1, LEM-2, and LEM-4 from further publication. Only LEM-3 and LEM-6 remain as active documents.

10.3 Energy Policy Act

On October 25, 1992, the Energy Policy Act was signed into law by the U.S. President. Among the many provisions, this Act establishes energy efficiency standards for HVAC, lighting, and motor equipment; encourages establishment of a national window energy efficiency rating system; and encourages state regulators to pursue demand-side-management (DSM) programs.

Under the bill, lighting manufacturers will have 3 years to stop making F96T12 and F96T12/HO 8-ft fluorescent lamps and some types of incandescent reflectors. Standard F40 lamps, except in the SP and SPX or equivalent types of high color-rendering lamps, would also fade away. General service incandescent lamps to be axed would include those from 30 to 100 W, in 115- to 130-V ratings, having medium screw bases, of both reflector and PAR types, having a diameter larger than 2 3/4 in.

There are no immediate regulations impacting HID lamps. Within 18 months of the legislation's enactment, the DOE will determine the HID types for which standards could possibly save energy and publish testing requirements for these lamps.

As far as the general service lamps are concerned, the most common incandescent lamps—40, 60, 75, 100, and 150 W—are not covered by an efficiency standard because there is no suitable method to ensure energy savings. These types, however, are covered by another provision of the law, namely, the energy efficiency labeling standards.

Table 10.1 The Proposed Efficiency Standards for Fluorescent Lamps

Lamp Type	Nominal lamp wattage	CRI	Minimum average lamp efficacy
F40	> 35 W	69	75
F40	</= 35 W	45	75
F40/U	> 35 W	69	68
F40/U	</= 35 W	45	64
F96T12	> 65 W	69	80
F96T12	</= 65 W	45	80
F96T12/HO	> 100 W	69	80
F96T12/HO	</= 100 W	45	80

Note: The above excludes lamps designed for plant growth, cold temperature service, reflectorized/aperture, impact resistance, reprographic service, colored lighting, ultraviolet, and lamps with CRI of more than 82.

Effective April 28, 1994, the Federal Trade Commission (FTC) must provide manufacturers with labeling requirements for all lamps covered: fluorescent, incandescent, and reflector incandescent. Though not yet defined, the proposals include: an energy rating for the lamps, probably LPW (lumens per watt), and energy cost per year is likely to be based on an operating cost of $0.10/kWh with 4-h/day operation. The energy efficiency label will then allow side-by-side comparison of two different lamp types, thus enabling consumers to make a more intelligent choice of lamps, taking into account not just the purchase price, but also the operating cost. Manufacturers must begin applying labels by April 28, 1995.

Table 10.1 shows the proposed efficiency standards for the fluorescent lamps, and Table 10.2 shows the proposed efficiency standards for incandescent reflector lamps.

Table 10.2 The Proposed Efficiency Standards for Incadescent Reflector Lamps

Nominal lamp wattage	Minimum average lamp efficacy (LPW)
40–50	10.5
51–66	11.0
67–85	12.5
86–115	14.0
116–155	14.5

Note: The above excludes miniature, decorative, traffic signal, marine, mine, stage/studio, railway, colored lamps, and other special application types.

There is no requirement to replace all existing lamps in any installation. However, as these lamps burn out, the replacement must meet the new standards. Replacement for popular fluorescent types includes reduced-wattage energy-saving types. These lamps will meet the color and efficiency standards; so will the full-wattage triphosphor lamps having a CRI over 69. On the incandescent side, replacements for the standard incandescent spot and flood lamps will be lower wattage halogen-type reflector lamps which do meet the LPW requirements. The halogen and halogen/infrared types of reflector lamps will remain the only type of such lamps on the market. ER and BR types, those intended for rough or vibration service, will also be excluded here.

In carrying out the replacement standards, expected savings on the lighting portion of this legislation will amount to 4300 MW in power and 37 billion kWh in energy usage. These huge savings are equal to the annual electric energy use of more than 4 million homes.

The earliest provision to the new law became effective at the end of April 1994. Effective that date, manufacturers could no longer produce 8-ft lamps which do not comply with the efficiency standards. Effective October 31, 1995, lamps covered included 4-ft fluorescents, 2-ft U-shaped fluorescents, and incandescent reflector types.

There is also a provision for lighting fixture manufacturers to come up with voluntary luminaire efficiency standards. If these standards are found to be inadequate, the Department of Energy will come up with the mandatory efficiency standards.

The new Energy Policy Act is all encompassing. It promises to change the way the industries produce, distribute, and utilize the valued energy resources.

10.4 U.S. state energy codes

State legislation governing lighting energy consumption varies from state to state. The federal Energy Policy Act of 1992 (EPAct) requires all states to adopt a building energy code that meets or exceeds the requirements of ASHRAE/IES 90.1-1989. The Model Energy Code (MEC) contains a codified version of standard 90.1-1989 and is used by many states to meet the EPAct requirement. The MEC has been developed by the Council of American Building Officials (CABO) with participation of the building code officials.

References

Lawrie, R. J., How the New Energy Law Will Affect You, EC&M, January 1993.

LEM-1, Lighting Power Limit Determination, Illuminating Engineering Society of North America, New York, 10005-4001.

LEM-3, Design Considerations for Effective Building Lighting Energy Utilization, Illuminating Engineering Society of North America, New York, 10005-4001.

Index

A

B

C

D

E

F

G

H

I

L

M

N

O

P

R

S

T

U

V

W

Z